KT-523-059

NIALL J.M.MILLER

JCT 80 and the Contractor

JCT 80 and the Contractor

I S Cherry, FRICS

Butterworths
London Boston Durban Singapore Sydney Toronto Wellington

All rights reserved. No part of this publication may be reproduced
or transmitted in any form or by any means, including
photocopying and recording, without the written permission of
the copyright holder, application for which should be addressed to
the Publishers. Such written permission must also be obtained
before any part of this publication is stored in a retrieval system of
any nature.

This book is sold subject to the Standard Conditions of Sale of
Net Books and may not be re-sold in the UK below the net price
given by the Publishers in their current price list.

First published 1985

© I. S. Cherry 1985

British Library Cataloguing in Publication Data

Cherry, I. S.
  JCT 80 and the Contractor.
  1. Joint Contracts Tribunal. Standard form of building
  contract. 1980   2. Building—Contracts and
  and specifications—Great Britain
  I. Title
  692′.8   TH425

  ISBN 0-408-01508-X

Photoset by Butterworths Litho Preparation Department
Printed and bound in Great Britain at the University Press, Cambridge

Preface

All opinions and conclusions in this book are my own, and for my part they reflect the meanings and intentions of the clauses in the Standard Forms. If, in some cases, I do not appear to be in complete accord with the Standard Form of Building Contract, this does not imply any criticism of the fine efforts which have gone into the preparation of the documents. Rather, that there are points to be noted which may promote discussion and therefore a better understanding. Though my opinions, at best, may be only academic, they will, I hope, add to the reader's appreciation of JCT 80.

The Contents list gives both the chapters and the clauses contained within them. A further list of clause numbers and headings is appended, these being referenced to the chapters in which they occur. Finally, there is a list of clauses which are changed by the Scottish Supplement and, again, these are referenced to the chapters in which they appear. It is hoped that this form of indexing will afford the reader an easy reference.

May I record here my appreciation of the kind permission to waive copyright received from the copyright holders of the following documents:

JCT 80 and associated tender documents	RIBA Publications Ltd
Extracts from NSC/4, NSC/4a and DOM/1	National Federation of Building Trades Employers
JCT80 and associated contract documents in Scotland	Building Employers' Confederation

without whose co-operation and permission to use this book would not have been possible.

I am also indebted to L. C. H. Bunton, FRICS, FCIArb, Chartered Quantity Surveyor and Arbiter of Glasgow, for his useful commentary on the text, and also for his help with the onerous task of proof reading.

I. S. Cherry FRICS

Contents

Index of clauses in chapters

England and Wales

Index of clauses

Amended by the Scottish Supplement for use in Scotland

Introduction

Within a series of four related books, this volume is designed to set out and explain, as simply as possible, one's involvement in a contract let under the Standard Form of Building Contract 1980 Edition.

Due to differences in Scottish law and traditional Scottish practice, the contract applicable in England and Wales requires minor modification and amendment to permit its operation in Scotland. Accordingly, the format of this book has been so selected as to allow for the differences between north and south of the Border.

For all the purposes of the conditions a contractor can be a person or a firm and is defined therein as:

England and Wales – The person named as a contractor in the Articles of Agreement
Scotland – The person named as a contractor in the Scottish building contract

In dealing with the individual clauses of the contract in detail, they are referred to by their number and heading and are not reproduced in their entirety. It is assumed that the printed documents are available within one's organization and are accessible for reference. Accordingly, the explanations, notes and diagrams are supplementary with a view to a simpler clarification of their meanings and intentions.

The book is laid out in chapters to group similar clauses under relevant headings and thus afford a ready reference.

Following each clause is the Scottish Supplement which adapts the clause for use in Scotland where it is relevant.

For immediate identification of the parts of the text and the contract forms relating to Scotland, these parts and forms have line drawn against them in the outside margin of each page. Also, the numbers of the Scottish *Diagrams* are followed by (S).

Published documents

England and Wales

The following list comprises documents published by the JCT in connection with the 1980 Edition of the Standard Form and Nominated Sub-contract procedures.

A   Standard Form of Building Contract 1980 Edition
1   Private With Quantities
2   Private Without Quantities
3   Private With Approximate Quantities
4   Local Authorities With Quantities
5   Local Authorities Without Quantities
6   Local Authorities With Approximate Quantities
7   Sectional Completion Supplement
NB All the standard forms above contain Articles of Agreement for the formation of the building contract.

B   Tender documents, etc. – Nominated Sub-contractors
1   NSC/1 – Nominated Sub-contract Tender and Agreement
2   NSC/2 – Employer/Nominated Sub-contractor Agreement for use where Sub-contractor tendered on NSC/1
3   NSC/2a – Employer/Nominated Sub-contractor Agreement for use where Tender NSC/1 not used
4   NSC/3 – Standard form of Nomination of Sub-contractor where Tender NSC/1 has been used

C   Standard Nominated Sub-contract
1   NSC/4 – Sub-contract for Sub-contractors who have tendered on Tender NSC/1 and executed Agreement NSC/2 and been nominated by Nomination NSC/3
2   NSC/4a – Sub-contract NSC/4 adapted for use where Tender NSC/1, Agreement NSC/2 and Nomination NSC/3 have not been used

D   Nominated Suppliers
1   TNS/1 – Standard Form of Tender for Nominated Supplier
2   TNS/2 – Warranty by a Nominated Supplier (to the Employer) (Schedule 3 to Tender TNS/1)

E   Other documents

1   Formula Rules for use with:

   (a)   Standard Form 1980 Edition With Quantities and With Approximate Quantities
   (b)   Standard Form 1980 Edition With Contractor's Design
   (c)   JCT Nominated Sub-contracts NSC/4 and NSC/4a

2   Fluctuation supplements for:

   (a)   Standard Form 1980 Edition Private and Local Authorities
   (b)   Standard Nominated Sub-contracts NSC/4 and NSC/4a

3   Standard Form of Contract With Contractor's Design
4   Contractor's Design portion supplement for use with Standard Form Private With Quantities and Local Authorities With Quantities
5   JCT Form of Tender by Nominated Supplier
6   JCT Guide to Standard Form of Building Contract, and the JCT Nominated Sub-contract Documents

The following documents are published by the NFBTE in connection with the 1980 Edition of the Standard Forms:

1   Sub-contract Conditions for use with the Domestic Sub-contract DOM/1 Articles of Agreement
2   Domestic Sub-contract DOM/1 Articles of Agreement

All documents are obtainable from:

RIBA Publications Ltd
   66 Portland Place, London W1N 4AD
National Federation of Building Trade Employers (NFBTE)
   82 New Cavendish Street, London W1M 8AD
Royal Institution of Chartered Surveyors (RICS)
   12 Great George Street, London SW1P 3AD

Scotland

The following list is of documents published by the JCT in connection with the 1980 Edition of the Standard Form and Nominated Sub-contract procedures which are incorporated by reference in the relevant Scottish documents.

A   Standard Form of Building Contract, 1980 Edition

1   Private With Quantities
2   Private Without Quantities
3   Private With Approximate Quantities
4   Local Authorities With Quantities
5   Local Authorities Without Quantities
6   Local Authorities With Approximate Quantities
7   Design and Build Contract

B   Nominated Sub-contract documents

1   NSC/4 – Standard Nominated Sub-contract 1980 for use when the Basic Method of nomination is employed
2   NSC/4a – Standard Nominated Sub-contract 1980 for use when the Alternative Method is employed

C   Other documents
1   Formula Rules for use with:

(a) Standard Form 1980 Edition with Quantities and with Approximate Quantities

2   Fluctuation supplements for:

(a) Standard Form 1980 Edition Private and Local Authorities
(b) Standard Nominated Sub-contracts NSC/4 and NSC/4a

List of documents published by the SBCC for use with the 1980 Edition of the Standard Form and Nominated Sub-contract procedures.

A   Main contract documents
1   Scottish Building Contract
2   Scottish Building Contract (Sectional Completion Supplement)
3   Scottish Building Contract with Approximate Quantities

B   Nominated Sub-contract documents
Documents used when the Basic Method of nomination is employed:

1   Nominated Sub-contract Tender NSC/1/Scot
2   Employer/Nominated Sub-contractor Agreement NSC/2/Scot
3   Standard Form of Nomination NSC/3/Scot
4   Scottish Building Sub-contract NSC/4/Scot

Documents used when the Alternative Method is employed:

5   Employer/Nominated Sub-contractor Agreement NSC/2a/Scot
6   Standard Form of Nomination NSC/3a/Scot
7   Scottish Building Sub-contract NSC/4a/Scot

C   Nominated Supplier documents
1   Standard Form of Tender by Nominated Supplier, including Warranty
2   Standard Form of Nomination of Nominated Suppliers

D   Other documents
Contract of purchase of off-site materials and goods
1   Property of Main Contractor
2   Property of Sub-contractor

Domestic Sub-contract documents:

3   Sub-contract Conditions for use with the Domestic Sub-contract DOM/1/Scot
4   Domestic Sub-contract DOM/1/Scot
5   Explanatory Memorandum to the 1980 Standard Form and relative Scottish documents

All documents are obtainable from:

Royal Incorporation of Architects in Scotland (RIAS)
15 Rutland Square, Edinburgh
Royal Institution of Chartered Surveyors (RICS), Scottish Branch
7 Manor Place, Edinburgh
Building Employers' Confederation (BEC)
13 Woodside Crescent, Glasgow

Notes on building contracts

The word 'contract' comes from the latin 'contractum' (drawn together) and denotes a drawing together of two or more minds to form a common intention, i.e. an agreement. Thus a contract is based upon an agreement of two or more parties.

The Law of Contract exists to protect such agreements, to ensure that parties to them carry out the various promises they have made, and to provide one party with a remedy should the other fail to do what he has undertaken to do.

Every contract is an agreement but not every agreement is a contract. In order for an agreement to be legally binding and an enforceable contract, three primary conditions must be fulfilled:

1   The agreement must relate to the future conduct of the parties. Two or more persons might agree about anything, e.g. that it is a wet day – but a mere consensus of opinion does not form a contract. The agreement must be to do or not to do something.
2   The parties must intend that their agreement shall be legally binding and enforceable – a mere moral obligation or a gentlemen's agreement is not a contract. Thus, if one party invites another to play golf and the other fails to turn up, the host cannot claim damages for breach of contract, for it cannot have been intended that their agreement was to have been legally binding.
3   The agreement must NOT infringe any rule or law. The law will not enforce an agreement designed to breach the law and so an illegal agreement is not a valid and enforceable contract.

Classification of contracts

Contracts are sub-divided into main groups:

(a) Contracts under seal
(b) Simple contracts

(a) *Contracts under seal* are written instruments which are signed and delivered by the parties. It is usual, although not essential, for the execution to have the signing attested by a witness and it must be stamped at the stamping office with a fee, if it is to be available as evidence before a court. The delivery is done by the party pronouncing such words as 'I deliver this as my act and deed' and handing the document to the other party. Nowadays these formalities seem to be dispensed with and all that is necessary is the signing of a written document, with a seal affixed, coupled with an intention to seal and deliver it. The document need not even be physically handed over.

Contracts under seal, either in writing or oral, are sometimes referred to collectively as *express contracts* in distinction from *implied contracts* where the agreement is implied from contract.

(NB These contracts under seal are made exclusively in England.)

(b) *Simple contracts* may be in writing (though writing is not generally necessary for the creation of a valid contract) or by spoken words, or as may be implied from the conduct of the parties. Examples of this are created every time a seat is taken on a bus or in a cinema without any words of agreement being spoken.

Looking now at the essential basis of a contract, as said previously, a contract is an agreement between two or more persons. It is usually found, on analysis, that agreement between two persons is the result of an offer made by one person and an acceptance of that offer by another. The person making the offer is known as the *offerer* and the person to whom it is made the *offeree*.

Constitution of an offer

An offer is constituted when there is a promise to undertake an obligation if the terms of the offer are accepted by the offeree. The offer must be such that mere acceptance of its terms by the offeree produces what can be seen to be agreement between the parties.

There are various actions which do not conform to the above definition of an offer and these must be distinguished from a true offer as under:

1   An offer to receive an offer, an invitation to treat
2   A declaration of intention, and
3   An offer which cannot be refused.

Termination of a contract

The various ways in which the contractual tie can be loosened and parties freed from their respective rights and liabilities under the contract, are as follows:

1   It may be performed, – i.e., the duties undertaken by both parties are fulfilled and all rights satisfied.
2   It may be discharged in the same way as it was created, i.e., by agreement.
3   It may be broken (or breached) by either one party or both parties.
4   It may be discharged by express provision contained in the contract.
5   It may become impossible of performance or be frustrated.

Of these five, number (3), *Breach of contract*, is the one most likely to be of interest to the Contractor, so the forms of breach will now be considered.

1   *Renunciation* occurs when one party states explicitly that he will not perform his promise.
2   *Implicit repudiation* occurs when one party does some act which disables him from performing his promise.
3   *Failure to perform* occurs when one party fails to perform his obligation on the date for its performance fixed by the contract.

Remedies for breach of contract

Damages

Whenever one party breaks a contract into which he has entered, the other party has the right of action agains him for damages. If the parties have agreed that in the event of a breach by one of them a certain sum shall become payable as damages, the sum agreed is known as *liquidated damages*. If, however, as is sometimes the case, the contract is silent as to any such sum, then the damages to be awarded are left to the decision of the court and are said to be *unliquidated*.

Liquidated damages

In the building contract and its terms it is important to know what is meant by 'liquidated damages'.

Where the parties to a contract agree that in the event of a breach the one shall pay the other a specified sum of money, that sum *may* be liquidated damages, but instead it may be what is called a *penalty*, and it is for the Court to decide which it is. The distinction is that, in their essence, liquidated damages represent a genuine and mutually-agreed estimate of the damages likely to result from a breach of the contract, but a penalty is essentially a payment of money stipulated for the intention of acting as a threat to secure performance and bears no relation to the actual damage which may be incurred.

If the stipulated sum is held to be liquidated damages then the plaintiff is entitled to recover that sum, no more and no less, whatever the actual damage he may have suffered. If, however, the stipulated sum is held to be a penalty then the plaintiff can recover in respect only of the damage he has suffered.

Where the penalty amount does not cover the damages suffered, the plaintiff may ignore the penalty and recover his damage in full. Thus a penalty has no effect as far as the recovery of damages is concerned.

Courts have laid down certain guiding rules as to whether or not a stipulated sum is a penalty, but they are not conclusive. If the sum is extravagant or inconscionable in comparison with the greatest loss there could be, it will be ruled as a penalty.

Unliquidated damages

The object of assessing and awarding unliquidated damages is to put the party who has suffered loss in as good a position, so far as money can achieve, as he would have been in had the contract been successfully performed.

There are other types of damages not usually associated with a building contract. Other remedies are:

Specific performance where a decree is issued by the Court compelling the defendant to carry out his promise to the plaintiff. This will be made only when the contract is certain, fair and just. The terms are usually difficult to supervise, and where the Court cannot supervise it is unlikely to grant such a decree.

Injunction is a Court order directing a party to undo or refrain from doing a particular act, and it may, in other cases, be an order where doing the act would constitute a breach of contract. An injunction may be issued where the Court cannot supervise a specific performance, as it virtually compels the party in default to perform his contract.

Quantum meruit: The right to *quantum meruit* ('so much as the thing is worth') is the right to reasonable remuneration for the work done. A claim for *quantum meruit* is based upon an implied contract between parties that a reasonable price be paid for work done. It follows, therefore, that no claim for a *quantum meruit* can succeed if it is inconsistent with the terms of an existing contract, if the complete performance is required before payment. But a party can treat a void contract as ended and sue on a *quantum meruit* for any work done under the void contract.

Chapter 1
Executing the formal contract with an employer

England and Wales

Each Standard Form of the 1980 Edition contains the Articles of Agreement at the beginning, the Appendix at the end, and the Conditions of the Contract in the middle, all of which, taken together, constitute the formal Contract.

It is usual in England and Wales to execute a formal contract. A sample Articles of Agreement and Appendix are incorporated in eight pages which have been filled up by an Employer to suit a contract let under the Private Edition With Quantities.

It is essential to scrutinize the documents very carefully, as they contain all the information with reference to alternatives to a number of clauses in the conditions and will affect the Contractor's pricing of his Preliminaries Bill.

Immediately following the sample documents there are explanatory notes.

It may be stating the obvious to say that there are only two parties to a building contract, namely the Employer and the Contractor. Domestic Sub-contractors have no involvement with the Employer and have only a normal contractual arrangement with the Contractor. Nominated Sub-contractors and Suppliers are on a different contractual basis which is discussed in full in a separate chapter.

The Joint Contracts Tribunal has issued the *JCT Guide* to the Standard Form of Building Contract 1980 Edition and to the JCT Nominated Sub-Contract Documents. Clauses 38, 39 and 40 referred to in clause 37 are issued separately for incorporation in this form. The Formula Rules referred to in clause 40 are also published separately in one booklet as are the Formula Rules for Sub-Contracts.

Variants of the Standard Form are:
Local Authorities
With Quantities
Without Quantities
With Approximate Quantities

Private
With Quantities
Without Quantities
With Approximate Quantities.

Sub-Contract documents also issued by the Tribunal for use with the Standard Form of Main Contract 1980 Edition are:
NSC/1 Tender
NSC/2 and 2a Employer/Sub-Contractor Agreement
NSC/3 Nomination of Sub-Contractor
NSC/4 and 4a JCT Nominated Sub-Contracts.

The variants of the Standard Form With Quantities and With Approximate Quantities may be adapted for use where the Works are to be completed in Phased Sections, by incorporating the Sectional Completion Supplement 1980 Edition.

Published by RIBA Publications Ltd
Finsbury Mission, Moreland Street
London EC1V 8VB

© RIBA Publications Ltd 1980

Reprinted May 1983
with corrections

Printed by Balding + Mansell Limited,
Wisbech, Cambs

Dated *6 June* 19 *82*

Standard Form of Building Contract
Private With Quantities
1980 Edition

Articles of Agreement and Conditions of Building Contract

between *The Capital Fund Building Society Capital House, 14-20 London Road, Birmingham*

and *Construction Services Ltd 181 Hangar Lane West Croydon, Surrey*

JCT

This Form is issued by the Joint Contracts Tribunal

Constituent bodies:
Royal Institute of British Architects
National Federation of Building Trades Employers
Royal Institution of Chartered Surveyors
Association of County Councils
Association of Metropolitan Authorities
Association of District Councils
Greater London Council
Committee of Associations of Specialist Engineering
   Contractors
Federation of Associations of Specialists and Sub-Contractors
Association of Consulting Engineers
Scottish Building Contract Committee

50p stamp
to be
impressed here
if contract is
under seal

Articles of Agreement

made the *Sixth* day of *June* 19 *82*

between *The Capital Fund Building Society*

of (or whose registered office is situated at) *Capital House,
14-20 London Road, Birmingham*

(hereinafter called 'the Employer') of the one part and

Construction Services Ltd

of (or whose registered office is situated at) *181 Hangar Lane
West Croydon, Surrey*

(hereinafter called 'the Contractor') [a] of the other part.

Whereas

Recitals First the Employer is desirous of [b] *the erection of a new four-storey
office block with all services and external
works complete*
(hereinafter called 'the Works') at

148 Mitcham Green, Mitcham, Surrey

and has caused Drawings and Bills of Quantities showing and describing the work to be done
to be prepared by or under the direction of

*Tripartite Design Group, 14A Bloomsbury Road,
London WC1*

Footnotes   [a] Where the Contractor is not a limited liability
company incorporated under the Companies Acts, see
Footnote [v] to clause 35 13 5 4 4

[b] State nature of intended works

Second   the Contractor has supplied the Employer with a fully priced copy of the said Bills of Quantities (which copy is hereinafter referred to as 'the Contract Bills');

Third   the said Drawings numbered *as listed in the Bill of Quantities*

(hereinafter referred to as 'the Contract Drawings') and the Contract Bills have been signed by or on behalf of the parties hereto;

Fourth   the status of the Employer, for the purposes of the statutory tax deduction scheme under the Finance (No. 2) Act, 1975, as at the Date of Tender is stated in the Appendix;

Now it is hereby agreed as follows

Contractor's obligations

Article 1
For the consideration hereinafter mentioned the Contractor will upon and subject to the Contract Documents carry out and complete the Works shown upon, described by or referred to in those Documents.

Contract sum

Article 2
The Employer will pay to the Contractor the sum of *one million, two hundred thousand and ten pounds - 39*

(£ *1 200 000 . 39*)

(hereinafter referred to as 'the Contract Sum') or such other sum as shall become payable hereunder at the times and in the manner specified in the Conditions.

Architect

Article 3
The term 'the Architect' in the Conditions shall mean the said

Tripartite Design Group

of *14th Bloomsbury Road*

London WC1

or, in the event of his death or ceasing to be the Architect for the purpose of this Contract, such other person as the Employer shall nominate for that purpose, not being a person to whom the Contractor shall object for reasons considered to be sufficient by an Arbitrator appointed in accordance with article 5. Provided always that no person subsequently appointed to be the Architect under this Contract shall be entitled to disregard or overrule any certificate or opinion or decision or approval or instruction given or expressed by the Architect for the time being.

Footnotes   [c] Not used.   [d] Not used.

P With 5/83

Quantity
Surveyor

Article 4
The Term 'the Quantity Surveyor' in the Conditions shall mean

Williams, Smith & Williams
Chartered Quantity Surveyors
of _400 St James Row._
London SE 1

or, in the event of his death or ceasing to be the Quantity Surveyor for the purpose of this
Contract, such other person as the Employer shall nominate for that purpose, not being a
person to whom the Contractor shall object for reasons considered to be sufficient by an
Arbitrator appointed in accordance with article 5.

Settlement of
disputes –
Arbitration

Article 5

5·1   In case any dispute or difference shall arise between the Employer or the Architect on his
behalf and the Contractor, either during the progress or after the completion or abandonment
of the Works, as to

5·1   ·1   the construction of this Contract, or

5·1   ·2   any matter or thing of whatsoever nature arising hereunder or in connection herewith
including any matter or thing left by this Contract to the discretion of the Architect or the
withholding by the Architect of any certificate to which the Contractor may claim to be
entitled or the adjustment of the Contract Sum under clause 30·6·2 or the rights and
liabilities of the parties under clauses 27, 28, 32 or 33 or unreasonable withholding of
consent or agreement by the Employer or the Architect on his behalf or by the Contractor,
but

5·1   ·3   excluding any dispute or difference under clause 31 to the extent provided in clause 31·9
and under clause 3 of the VAT Agreement,

then such dispute or difference shall be and is hereby referred to the arbitration and final
decision of a person to be agreed between the parties to act as Arbitrator, or, failing agreement
within 14 days after either party has given to the other a written request to concur in the
appointment of an Arbitrator, a person to be appointed on the request of either party by the
President or a Vice-President for the time being of the Royal Institute of British Architects.

5·1   ·4   Provided that if the dispute or difference to be referred to arbitration under this Contract
raises issues which are substantially the same as or connected with issues raised in a
related dispute between

the Employer and a Nominated Sub-Contractor under Agreement NSC/2 or NSC/2a
as applicable or

the Contractor and any Nominated Sub-Contractor under Sub-Contract NSC/4 or
NSC/4a as applicable or

the Contractor and/or the Employer and any Nominated Supplier whose contract of
sale with the Contractor provides for the matters referred to in clause 36·4·8,

and if the related dispute has already been referred for determination to an Arbitrator, the
Employer and Contractor hereby agree that the dispute or difference under this Contract
shall be referred to the Arbitrator appointed to determine the related dispute; and such
Arbitrator shall have power to make such directions and all necessary awards in the same
way as if the procedure of the High Court as to joining one or more defendants or joining
co-defendants or third parties was available to the parties and to him;

Footnote   [e] Not used.

5·1 ·5 save that the Employer or the Contractor may require the dispute or difference under this Contract to be referred to a different Arbitrator (to be appointed under this Contract) if either of them reasonably considers that the Arbitrator appointed to determine the related dispute is not appropriately qualified to determine the dispute or difference under this Contract.

5·1 ·6 Articles 5·1·4 and 5·1·5 shall apply unless in the Appendix the words "Articles 5·1·4 and 5·1·5 apply" have been deleted.

5·2 Such reference, except

·1 on article 3 or article 4; or

·2 on the questions
whether or not the issue of an instruction is empowered by the Conditions; or
whether or not a certificate has been improperly withheld; or
whether a certificate is not in accordance with the Conditions; or

·3 on any dispute or difference under clause 4·1 in regard to a reasonable objection by the Contractor, and clauses 25, 32 and 33,

shall not be opened until after Practical Completion or alleged Practical Completion of the Works or termination or alleged termination of the Contractor's employment under this Contract or abandonment of the Works, unless with the written consent of the Employer or the Architect on his behalf and the Contractor.

5·3 Subject to the provisions of clauses 4·2, 30·9, 38·4·3, 39·5·3 and 40·5 the Arbitrator shall, without prejudice to the generality of his powers, have power to direct such measurements and/or valuations as may in his opinion be desirable in order to determine the rights of the parties and to ascertain and award any sum which ought to have been the subject of or included in any certificate and to open up, review and revise any certificate, opinion, decision, requirement or notice and to determine all matters in dispute which shall be submitted to him in the same manner as if no such certificate, opinion, decision, requirement or notice had been given.

5·4 The award of such Arbitrator shall be final and binding on the parties.

5·5 Whatever the nationality, residence or domicile of the Employer, the Contractor, any sub-contractor or supplier or the Arbitrator, and wherever the Works or any part thereof are situated, the law of England shall be the proper law of this Contract and in particular (but not so as to derogate from the generality of the foregoing) the provisions of the Arbitration Acts 1950 (notwithstanding anything in S.34 thereof) to 1979 shall apply to any arbitration under this Contract wherever the same, or any part of it, shall be conducted.[f]

Footnote   [f] Where the parties do not wish the proper law of the Contract to be the law of England and/or do not wish the provisions of the Arbitration Acts 1950 to 1979 to apply to any arbitration under the Contract held under the procedural law of Scotland (or other country) appropriate amendments to article 5·5 should be made

Signed by or on behalf of the Employer [g1] _____

    in the presence of :

Signed by or on behalf of the Contractor [g1] _____

    in the presence of :

Signed, sealed and delivered by [g2]/The common seal of [g3] : _____

    in the presence of [g2]/was hereunto affixed in the presence of [g3] :

Signed, sealed and delivered by [g2]/The common seal of [g3] : _____

    in the presence of [g2]/was hereunto affixed in the presence of [g3] :

Footnotes

[g1] For use if Agreement is executed under hand.

[g2] For use if executed under seal by an individual or firm or unincorporated body.

[g3] For use if executed under seal by a company or other body corporate.

Appendix

	Clause	
Statutory tax deduction scheme – Finance (No. 2) Act 1975	Fourth recital and 31	Employer at Date of Tender *is a [contractor]* is not a 'contractor' for the purposes of the Act and the Regulations *(Delete as applicable)
Settlement of disputes – Arbitration	5·1	Articles 5·1·4 and 5·1·5 apply (See Article 5·1·6)
Date for Completion	1·3	*5th August 1984*
Defects Liability Period (if none other stated is 6 months from the day named in the Certificate of Practical Completion of the Works)	17·2	*12 months*
Insurance cover for any one occurrence or series of occurrences arising out of one event	21·1·1	£ *5 000 000*
Percentage to cover professional fees	22A	*N/A*
Date of Possession	23·1	*4th August 1982*
liquidated and ascertained damages	24·2	at the rate of £ *2000* per *week*
Period of delay: [z]	28·1·3	
(i) by reason of loss or damage caused by any one of the Clause 22 Perils	28·1·3·2	*3 months*
(ii) for any other reason	28·1·3·1, 28·1·3·3 to ·3·7	*1 month*
Period of Interim Certificates (if none stated is one month)	30·1·3	*1 month*
Retention Percentage (if less than 5 per cent) [aa]	30·4·1·1	*3%*
Period of Final Measurement and Valuation (if none stated is 6 months from the day named in the Certificate of Practical Completion of the Works)	30·6·1·2	*12 months*

Footnotes

[z] It is suggested that the periods should respectively be three months and one month. It is essential that periods be inserted since otherwise no period of delay would be prescribed.

[aa] The percentage will be 5 per cent unless a lower rate is specified here.

Work reserved for Nominated Sub-Contractors for which the Contractor desires to tender	35·2	————————————
Fluctuations: (if alternative required is not shown clause 38 shall apply)	37	~~clause 38~~ [cc] clause 39 ~~clause 40~~
Percentage addition	~~38·7 or~~ 39·8	*5%* ————————
Formula Rules	40·1·1·1	—
	rule 3	Base Month ——— 19 —
	rules 10 and 30 (i)	~~(Part I/Part II [dd] of Section 2 of the Formula Rules is to apply)~~

Footnotes

[bb] Not used.

[cc] Delete alternatives not used.

[dd] Strike out according to which method of formula adjustment (Part I – Work Category Method or Part II – Work Group Method) has been stated in the Bills of Quantities issued to tenderers.

Explanatory notes

Articles of Agreement	A 50p stamp should be affixed at the top of Page 5 if the Contract is under seal.
Third Recital	Drawing numbers may be either listed here or referred to a list in the Bill of Quantities.
Article Five	Note the reference under Clause 5.1.6 to the Appendix where an alternative may be deleted.
Attestation	Note here the different forms of agreement. Agreement under hand, under seal, by an individual or firm or unincorporated body, or under seal by a company or other body corporate.

Appendix

Fourth Recital and Clause 31	The status of the Employer should be ascertained here by striking out an alternative.
Article 5.1	The words 'Articles 5.1.4 and 5.1.5 apply' should be either left in or struck out.
Clause 37	Only one clause should remain.
Clause 38.7 or 39.8	Only one of the two clauses should apply.
Clause 40.1.1.1.	Will be filled in only if Clause 40 is operative.

Sectional Completion Supplement

This is the formal Contract between the Employer and the Contractor where sectional completion has been agreed at the tender stage.

A sample form follows which has been filled up by an Employer with the necessary deletions and amendments to suit a contract let under the Private Edition with Quantities.

It is important to check the form with regard to any deletions and changes made and that it agrees with the Appendix in the Preliminaries Bill. The Appendix this time is split into columns to represent the number of sections under which the works have to be carried out and these sections are numbered or otherwise identified.

Following are notes on the pages where changes or deletions have taken place and the sample form contains Notes on the Supplement and Practice Note 1 at the end. The numbering of the clauses refers to the Standard Conditions of Contract.

Apart from the Bill of Quantities where the layout of the Bill will be in the various sections of the Work, the Sectional Completion Supplement is the only document where the changes to the relevant clauses are recorded and is so incorporated into the Contract.

Note that Clause 18 – Partial possession by the Employer – is not intended as an alternative to the Sectional Completion Supplement. Where this form has been used the sections have been pre-determined at tender stage. Clause 18 is used where possession of part of the works was not planned at tender stage and was not an obligation imposed by the Employer.

The Joint Contracts Tribunal has issued the *JCT Guide* to the Standard Form of Building Contract 1980 Edition and to the JCT Nominated Sub-Contract Documents. Clauses 38, 39 and 40 referred to in clause 37 are issued separately for incorporation in this form. The Formula Rules referred to in clause 40 are also published separately in one booklet as are the Formula Rules for Sub-Contracts.

Variants of the Standard Form are:
Local Authorities
With Quantities
Without Quantities
With Approximate Quantities

Private
With Quantities
Without Quantities
With Approximate Quantities.

Sub-Contract documents also issued by the Tribunal for use with the Standard Form of Main Contract 1980 Edition are:
NSC/1 Tender
NSC/2 and 2a Employer/Sub-Contractor Agreement
NSC/3 Nomination of Sub-Contractor
NSC/4 and 4a JCT Nominated Sub-Contracts.

The variants of the Standard Form With Quantities and With Approximate Quantities may be adapted for use where the Works are to be completed in Phased Sections, by incorporating the Sectional Completion Supplement 1980 Edition.

Dated 6 June 19 82

Standard Form of Building Contract
Private With Quantities
1980 Edition

Articles of Agreement and Conditions of Building Contract

between *the Capital Fund Building Society Capital House, 14-20 London Road, Birmingham*

and *Construction Services Ltd 181 Hangar Lane West Croydon, Surrey*

JCT

This Form is issued by the Joint Contracts Tribunal

Constituent bodies:
Royal Institute of British Architects
National Federation of Building Trades Employers
Royal Institution of Chartered Surveyors
Association of County Councils
Association of Metropolitan Authorities
Association of District Councils
Greater London Council
Committee of Associations of Specialist Engineering
    Contractors
Federation of Associations of Specialists and Sub-Contractors
Association of Consulting Engineers
Scottish Building Contract Committee

Published by RIBA Publications Ltd
Finsbury Mission, Moreland Street
London EC1V 8VB

© RIBA Publications Ltd 1980

Reprinted May 1983
with corrections

Printed by Balding + Mansell Limited,
Wisbech, Cambs

12

50p stamp
to be
impressed here
if contract is
under seal

Articles of Agreement

made the _Sixth_ day of _June_ 19 _82_

between _The Capital Fund Building Society_

of (or whose registered office is situated at) _Capital House,_
14-20 London Road, Birmingham

(hereinafter called 'the Employer') of the one part and

Construction Services Ltd

of (or whose registered office is situated at) _181 Hangar Lane_
West Croydon, Surrey

(hereinafter called 'the Contractor') [a] of the other part.

Whereas

Recitals

First the Employer is desirous of [b] _the erection of a new four-storey_
office block with all services and external
works complete.
(hereinafter called 'the Works') at

148 Mitcham Green, Mitcham, Surrey

and has caused Drawings and Bills of Quantities showing and describing the work to be done
to be prepared by or under the direction of

Tripartite Design Group, 14A Bloomsbury Road,
London WC1

Footnotes

[a] Where the Contractor is not a limited liability
company incorporated under the Companies Acts, see
Footnote [v] to clause 35·13·5·4·4.

[b] State nature of intended works.

Second the Contractor has supplied the Employer with a fully priced copy of the said Bills of Quantities (which copy is hereinafter referred to as 'the Contract Bills');

Third the said Drawings numbered *as listed in the Bill of Quantities*

(hereinafter referred to as 'the Contract Drawings') and the Contract Bills have been signed by or on behalf of the parties hereto;

Fourth the status of the Employer, for the purposes of the statutory tax deduction scheme under the Finance (No. 2) Act, 1975, as at the Date of Tender is stated in the Appendix;

Now it is hereby agreed as follows

Contractor's obligations

Article 1
For the consideration hereinafter mentioned the Contractor will upon and subject to the Contract Documents carry out and complete the Works shown upon, described by or referred to in those Documents.

Contract sum

Article 2
The Employer will pay to the Contractor the sum of *one million, two hundred thousand and ten pounds · 39*

(£ *1 200 000 · 39*)

(hereinafter referred to as 'the Contract Sum') or such other sum as shall become payable hereunder at the times and in the manner specified in the Conditions.

Architect

Article 3
The term 'the Architect' in the Conditions shall mean the said

Tripartite Design Group

of *14ⁿ Bloomsbury Road*

London WC1

or, in the event of his death or ceasing to be the Architect for the purpose of this Contract, such other person as the Employer shall nominate for that purpose, not being a person to whom the Contractor shall object for reasons considered to be sufficient by an Arbitrator appointed in accordance with article 5. Provided always that no person subsequently appointed to be the Architect under this Contract shall be entitled to disregard or overrule any certificate or opinion or decision or approval or instruction given or expressed by the Architect for the time being.

Footnotes　　　　[c] Not used.　　　　　　　　　　[d] Not used.

14

Quantity
Surveyor

Article 4

The Term 'the Quantity Surveyor' in the Conditions shall mean

Williams, Smith & Williams
Chartered Quantity Surveyors

of *400 St James Row*

London SE1

or, in the event of his death or ceasing to be the Quantity Surveyor for the purpose of this Contract, such other person as the Employer shall nominate for that purpose, not being a person to whom the Contractor shall object for reasons considered to be sufficient by an Arbitrator appointed in accordance with article 5.

Settlement of
disputes –
Arbitration

Article 5

5·1   In case any dispute or difference shall arise between the Employer or the Architect on his behalf and the Contractor, either during the progress or after the completion or abandonment of the Works, as to

5·1   ·1   the construction of this Contract, or

5·1   ·2   any matter or thing of whatsoever nature arising hereunder or in connection herewith including any matter or thing left by this Contract to the discretion of the Architect or the withholding by the Architect of any certificate to which the Contractor may claim to be entitled or the adjustment of the Contract Sum under clause 30·6·2 or the rights and liabilities of the parties under clauses 27, 28, 32 or 33 or unreasonable withholding of consent or agreement by the Employer or the Architect on his behalf or by the Contractor, but

5·1   ·3   excluding any dispute or difference under clause 31 to the extent provided in clause 31·9 and under clause 3 of the VAT Agreement,

then such dispute or difference shall be and is hereby referred to the arbitration and final decision of a person to be agreed between the parties to act as Arbitrator, or, failing agreement within 14 days after either party has given to the other a written request to concur in the appointment of an Arbitrator, a person to be appointed on the request of either party by the President or a Vice-President for the time being of the Royal Institute of British Architects.

5·1   ·4   Provided that if the dispute or difference to be referred to arbitration under this Contract raises issues which are substantially the same as or connected with issues raised in a related dispute between

the Employer and a Nominated Sub-Contractor under Agreement NSC/2 or NSC/2a as applicable or

the Contractor and any Nominated Sub-Contractor under Sub-Contract NSC/4 or NSC/4a as applicable or

the Contractor and/or the Employer and any Nominated Supplier whose contract of sale with the Contractor provides for the matters referred to in clause 36·4·8,

and if the related dispute has already been referred for determination to an Arbitrator, the Employer and Contractor hereby agree that the dispute or difference under this Contract shall be referred to the Arbitrator appointed to determine the related dispute; and such Arbitrator shall have power to make such directions and all necessary awards in the same way as if the procedure of the High Court as to joining one or more defendants or joining co-defendants or third parties was available to the parties and to him;

Footnote   [e] Not used.

5·1 ·5 save that the Employer or the Contractor may require the dispute or difference under this Contract to be referred to a different Arbitrator (to be appointed under this Contract) if either of them reasonably considers that the Arbitrator appointed to determine the related dispute is not appropriately qualified to determine the dispute or difference under this Contract.

5·1 ·6 Articles 5·1·4 and 5·1·5 shall apply unless in the Appendix the words "Articles 5·1·4 and 5·1·5 apply" have been deleted.

5·2 Such reference, except

·1 on article 3 or article 4; or

·2 on the questions
whether or not the issue of an instruction is empowered by the Conditions; or
whether or not a certificate has been improperly withheld; or
whether a certificate is not in accordance with the Conditions; or

·3 on any dispute or difference under clause 4·1 in regard to a reasonable objection by the Contractor, and clauses 25, 32 and 33,

shall not be opened until after Practical Completion or alleged Practical Completion of the Works or termination or alleged termination of the Contractor's employment under this Contract or abandonment of the Works, unless with the written consent of the Employer or the Architect on his behalf and the Contractor.

5·3 Subject to the provisions of clauses 4·2, 30·9, 38·4·3, 39·5·3 and 40·5 the Arbitrator shall, without prejudice to the generality of his powers, have power to direct such measurements and/or valuations as may in his opinion be desirable in order to determine the rights of the parties and to ascertain and award any sum which ought to have been the subject of or included in any certificate and to open up, review and revise any certificate, opinion, decision, requirement or notice and to determine all matters in dispute which shall be submitted to him in the same manner as if no such certificate, opinion, decision, requirement or notice had been given.

5·4 The award of such Arbitrator shall be final and binding on the parties.

5·5 Whatever the nationality, residence or domicile of the Employer, the Contractor, any sub-contractor or supplier or the Arbitrator, and wherever the Works or any part thereof are situated, the law of England shall be the proper law of this Contract and in particular (but not so as to derogate from the generality of the foregoing) the provisions of the Arbitration Acts 1950 (notwithstanding anything in S.34 thereof) to 1979 shall apply to any arbitration under this Contract wherever the same, or any part of it, shall be conducted.[f]

Footnote [f] Where the parties do not wish the proper law of the Contract to be the law of England and/or do not wish the provisions of the Arbitration Acts 1950 to 1979 to apply to any arbitration under the Contract held under the procedural law of Scotland (or other country) appropriate amendments to article 5·5 should be made

16

Signed by or on behalf of the Employer [g1] _____

    in the presence of:

Signed by or on behalf of the Contractor [g1] _____

    in the presence of:

Signed, sealed and delivered by [g2]/The common seal of [g3]: _____

    in the presence of [g2]/was hereunto affixed in the presence of [g3]:

Signed, sealed and delivered by [g2]/The common seal of [g3]: _____

    in the presence of [g2]/was hereunto affixed in the presence of [g3]:

Footnotes

[g1] For use if Agreement is executed under hand.

[g2] For use if executed under seal by an individual or firm or unincorporated body.

[g3] For use if executed under seal by a company or other body corporate.

Article 6

The modifications to the Form of Agreement stated in the Table set out below are hereby incorporated in this Contract and the provisions of the Articles of Agreement and the annexed Conditions shall have effect as so modified:

TABLE OF CHANGES TO THE ARTICLES OF AGREEMENT AND THE ANNEXED CONDITIONS [1]

The clauses indicated in the first column of this Table are modified, at the places therein shown in the second column, by the deletion or insertion of such words as are shown in the third column.

1	2	3
Clause number	Place in text	Words to be deleted/inserted
Article 5·2	AT END of Article 5·2	ADD 'or unless in the Appendix it is agreed that a reference to arbitration may be opened on practical completion of any relevant Section'.
Article 5·5	AFTER 'wherever the Works, or any'	INSERT 'Section or'
1·3	AT END of definition of 'Conditions'	INSERT 'together with the modifications referred to in Article 6.'
	BETWEEN the definitions of 'Retention Percentage' and 'Statutory Requirements'	INSERT as an additional definition: 'Section: one of the Sections into which the Works have been divided for phased completion as shown upon the Contract Drawings and described by or referred to in the Contract Bills and in the Appendix.'
	In the definition of 'Completion Date' AFTER 'Completion'	INSERT 'for each Section'
	In the definition of 'Date for Completion' AFTER 'stated'	INSERT 'for each Section'
	In the definition of 'Date of Possession' AFTER 'stated'	INSERT 'for each Section'
	In the definition of 'Defects Liability Period' AFTER 'named'	INSERT 'for each Section'
2·1	AFTER 'the Contractor shall upon and subject to the Conditions carry out and complete the Works'	INSERT 'by Sections'
17·1	AFTER 'When in the opinion of the Architect/Supervising Officer Practical Completion of'	DELETE 'the Works' AND INSERT 'any Section'
	AFTER 'effect and'	DELETE 'Practical Completion of the Works' and INSERT 'practical completion of that Section'
17·2	BEFORE 'within the Defects Liability Period'	INSERT 'in any Section'
	AFTER these words	INSERT 'in relation thereto'
	AFTER 'occurring before'	DELETE 'Practical Completion of' INSERT 'practical completion of that Section of'
	AFTER 'specified by the Architect/Supervising Officer in a schedule of defects'	INSERT 'for that Section'
17·3	BEFORE 'within the Defects Liability Period'	INSERT 'in any Section'
	AFTER these words	INSERT 'in relation thereto'
	AFTER 'frost occurring before'	DELETE 'Practical Completion of the Works' and INSERT 'practical completion of that Section'

Footnote	[1] The words 'Supervising Officer' do not appear in the Private edition and for Private Edition contracts should be struck out or be deemed to be deleted wherever	appearing in this Table. In the Local Authorities Edition the term 'Supervising Officer' is substituted throughout where the official concerned is not an Architect.
Source	Standard Form of Sectional Completion Supplement issued by the Joint Contracts Tribunal for use with the Standard Form of Building Contract WITH Quantities or WITH APPROXIMATE Quantities, Private and Local Authorities editions. 1980. See Notes C and D on page 2 of Supplement.	

1	2	3
Clause number	**Place in text**	**Words to be deleted/inserted**
17·4	AFTER 'and completion of making good defects'	INSERT 'in the relevant Section'
17.5	AFTER 'appear after'	DELETE 'Practical Completion' INSERT 'practical completion of any Section'
	AFTER 'which took place before'	DELETE 'Practical Completion' INSERT 'practical completion of that Section'
17	AFTER clause 17·5	INSERT as new clause 17 6: 'When in the opinion of the Architect/Supervising Officer practical completion of all the Sections has been achieved he shall forthwith issue (in addition to any certificates of practical completion of the Sections) a Certificate of Practical Completion of the Works and practical completion of the whole of the Works shall for the purpose of clause 30·6·1·1 be deemed to have taken place on the day named in such Certificate.'
18·1	AFTER 'If at any time or times before'	DELETE 'Practical Completion of the Works' and INSERT 'practical completion of any Section'
18·1·5	AFTER 'any period during which'	DELETE 'the Works' and INSERT 'any Section'
	AFTER 'provisions of clause 18 as does the'	DELETE 'Contract Sum' [2] and INSERT 'Section value'
	AFTER 'the said relevant part to the'	DELETE 'Contract Sum' [2] and INSERT 'Section value'
	AFTER clause 18·1·5	INSERT new clause 18·1·6: 'For the purposes of clause 18 the expression 'Section value' shall mean the value ascribed to the relevant Section in the Appendix'
22A	AFTER clause 22A·3·2	INSERT as additional clause 22A·3·3; 'Upon practical completion of any Section that Section shall be at the sole risk of the Employer as regards any of the Clause 22 Perils and after taking into account any operation of the provisions of clause 18·1·4 the Contractor shall reduce the value insured under clause 22A·1 by the value ascribed to that Section in the Appendix or such other value as may be specifically agreed in writing and recorded by amendment of the relevant part of the Appendix.'
23·1	BEFORE 'possession of the'	INSERT 'in relation to any Section'
	AFTER 'possession of the'	INSERT 'relevant part of the'
	AFTER ' shall thereupon begin'	DELETE 'the Works' and INSERT 'that Section'
	AFTER 'Completion Date'	INSERT 'in relation thereto'
24·1	AFTER 'if the Contractor fails to complete'	DELETE 'the Works' and INSERT 'any Section'
	AFTER 'by the Completion Date'	INSERT 'in relation thereto'
24·2·1	AFTER 'calculated at the rate stated'	INSERT 'in relation thereto'
	AFTER 'Completion Date'	INSERT 'of that Section'
	AFTER 'and the date of'	DELETE 'Practical Completion' INSERT 'practical completion in relation thereto'

Footnote   [2] In Approximate Quantities variant 'Tender Price'.

1	2	3
Clause number	Place in text	Words to be deleted/inserted
24·2·2	AFTER 'fixes a later Completion Date'	INSERT 'for any Section'
	AFTER 'such later Completion Date'	INSERT 'in relation to that Section'
25·2·1·1	AFTER 'the progress of'	DELETE 'the Works' and INSERT 'any Section'
25·2·2·2	AFTER 'in the completion of'	DELETE 'the Works' and INSERT 'that Section'
	AFTER 'the Completion Date'	INSERT 'in relation thereto'
25·3·1	AFTER 'date as the Completion Date'	INSERT 'for any such Section'
25·3·1·2	AFTER 'the completion of'	DELETE 'the Works' and INSERT 'any Section or Sections'
	AFTER 'the Completion Date'	INSERT 'in relation thereto'
25·3·2	AFTER 'clause 25·3·1'	INSERT 'in relation to any Section'
	AFTER 'fix a Completion Date'	INSERT 'for that Section'
	AFTER 'the omission'	INSERT 'from that Section'
	AFTER 'extension of time'	INSERT 'for that Section'
25·3·3	AFTER 'date of'	DELETE 'Practical Completion' INSERT 'practical completion of any Section'
25·3·3·1	AFTER 'fix a Completion Date'	INSERT 'for that Section'
25·3·3·2	AFTER 'fix a Completion Date'	INSERT 'for that Section'
	AFTER 'the omission'	INSERT 'from that Section'
	AFTER 'extension of time'	INSERT 'for that Section'
25·3·4·1	AFTER 'progress of'	DELETE 'the Works' and INSERT 'any Section'
	AFTER 'completion of'	DELETE 'the Works' and INSERT 'any Section'
	AFTER 'Completion Date'	ADD 'in relation thereto'
25·3·4·2	AFTER 'Works'	ADD 'or any Section thereof'
25·3·5	AFTER 'Completion Date'	ADD 'for any Section'
25·3·6	AFTER 'Completion Date' AFTER 'Date for Completion'	INSERT 'for any Section' INSERT 'in relation thereto'
25·4·6	AFTER 'Completion Date'	INSERT 'for any Section'
26·1	AFTER 'regular progress of the Works or of any'	INSERT 'Section or'
26·1·1	AFTER 'regular progress of the Works or of any'	INSERT 'Section or'
26·4·1	AFTER 'regular progress of the Sub-Contract works or of any'	INSERT 'Section or'
29·1	AFTER 'Complete the Works'	INSERT 'or any Section thereof'
30·1·3	AFTER 'Certificate of Practical Completion'	INSERT 'of the whole of the Works under Clause 17·6'
30·6·1·1	BEFORE 'Practical Completion of the Works'	INSERT 'the issue by the Architect/Supervising Officer of the Certificate of'
	AFTER these words	INSERT 'as required by Clause 17·6 (or where relevant the certificate of practical completion of any Section of the Works).'

1		2	3
Clause number		Place in text	Words to be deleted/inserted
30·6·1·2		AFTER 'within the Period'	INSERT '(or Periods, if Final Measurement and Valuation is to be completed in respect of each Section).'
30·8	*Local Authorities only* [3]	AFTER 'Before the expiration of'	DELETE 'the period the length of which is stated in the Appendix' and INSERT '3 months'
		AFTER 'Defects Liability Period'	DELETE 'also'
		AFTER 'in the Appendix'	INSERT 'in respect of the Section last completed'
		AFTER 'Clause 17'	INSERT 'in respect of all Sections'
32·1·2		AFTER 'the Works or any'	INSERT 'Section or'
33·1		AFTER 'In the event of the Works or any'	INSERT 'Section or'
33·3		AFTER 'sustained by the Works or any'	INSERT 'Section or'
38·4·7	*Not applicable to Form 'with Approximate Quantities'*	AFTER 'or 38.3'	INSERT 'for any Section'
		AFTER 'Completion Date'	INSERT 'in relation thereto'
38·4·8		AFTER 'not be applied'	INSERT 'in relation to any Section'
38·4·8·1		AFTER 'clause 25'	INSERT 'as hereinbefore modified'
38·4·8·2		AFTER 'Completion Date'	INSERT 'for that Section'
39·5·4		AFTER 'or clause 39·4'	INSERT 'for any Section'
39·5·8		AFTER 'not be applied'	INSERT 'in relation to any Section'
39·5·8·1		AFTER 'clause 25'	INSERT 'as hereinbefore modified'
39·5·8·2		AFTER 'Completion Date'	INSERT 'for that Section'
40·7·1·1		AFTER 'if the Contractor fails to complete'	DELETE 'the Works' and INSERT 'any Section'
		AFTER 'by the Completion Date'	INSERT 'in relation thereto'
		AFTER 'under clause 40'	INSERT 'relevant to that Section'
40·7·2		AFTER 'not be applied'	INSERT 'in relation to any Section'
40·7·2·1		AFTER 'clause 25'	INSERT 'as hereinbefore modified'
40·7·2·2		AFTER 'Completion Date'	INSERT 'for that Section'
Appendix		AFTER the end of the Conditions	DELETE the form of Appendix and INSERT an Appendix drawn up as shown in the Standard Form Sectional Completion Supplement from which this Table is extracted.

Footnote   [3] Delete where Private Edition applies.

T

Appendix (Sectional Completion Supplement)

	Clause	
Statutory tax deduction scheme – Finance (No. 2) Act 1975	Fourth recital and 31	Employer at Date of Tender [*] is a (contractor) is not a 'contractor' for the purposes of the Act and the Regulations
Settlement of disputes – Arbitration	Article 5·1	Articles 5·1·4 and 5·1·5 apply (See Article 5·1·6)
Time for commencement of Arbitration	Article 5·2	After Practical Completion of Works Relevant Section [*]

		Section Number	Section Number	Section Number
		1	*2*	*3*
Section of Works as shown on the Contract Drawings and described in the Contract Bills	2·1			
Section value (total value of Section ascertained from Contract Bills)	18·1·5, 22A	£800.000.37	£20.005.00	£20.005.00
Defects Liability period (if none stated is 6 months from the day named in the certificate of practical completion of Section)	17, 18, 30	12 MONTHS	12 MONTHS	12 MONTHS
Date of Possession of Section	23·1	4/8/82	5/8/83	5/8/83
Date for Completion of Section	1·3	5/8/83	5/8/84	5/8/84
Rate of Liquidated and Ascertained damages for Section	24·2·1	£1500 per WEEK	£750. per WEEK	£750. per WEEK
Insurance cover for any one occurrence or series of occurrences arising out of one event	21·1·1	£ 5 000 000		
Percentage to cover professional fees	22A	N/A		
Period of delay: [z]	28·1·3			
(i) by reason of loss or damage caused by any one of the Clause 22 Perils	28·1·3·2	3 MONTHS		
(ii) for any other reason	28·1·3·1, 28·1·3·3. to ·3·7	1 MONTH		
Period of Interim Certificates (if none stated is one month)	30·1·3	1 MONTH		
Retention Percentage (if less than 5 per cent) [aa]	30·4·1·1	3 %		
Period of Final Measurement and Valuation (if none stated is 6 months from the day named in the Certificate of Practical Completion of the Works or of any Section where so agreed)	30·6·1·2	12 MONTHS		

Footnotes

[*] Delete as applicable

[z] It is suggested that the periods should respectively be three months and one month. It is essential that periods be inserted since otherwise no period of delay would be prescribed.

[aa] The percentage will be 5 per cent unless a lower rate is specified here.

[bb] Not used.

Source Standard Form of Sectional Completion Supplement issued by the Joint Contracts Tribunal for use with the Standard Form of Building Contract WITH Quantities or WITH APPROXIMATE Quantities, Private and Local Authorities editions, 1980. See Note D on page 2 of Supplement.

Work reserved for Nominated Sub-Contractors for which the Contractor desires to tender	35·2	*JOINERY FITTINGS*
Fluctuations: (if alternative required is not shown clause 38 shall apply) [4]	37	~~clause 38 [4] [cc]~~ clause 39 ~~clause 40~~
Percentage addition	~~38·7 or~~ 39·8	*5%*
Formula Rules	40·1·1·1	—
	rule 3	Base Month _____ 19 ___
	rule 3	~~Non-Adjustable Element~~ ~~(not to exceed 10%) [5]~~
	rules 10 and 30 (i)	~~Part I/Part II [dd] of Section 2 of the Formula Rules is to apply~~

Footnotes

[cc] Delete alternatives not used.

[dd] Strike out according to which method of formula adjustment (Part I – Work Category Method or Part II – Work Group Method) has been stated in the Bills of Quantities issued to tenderers.

[4] Clause 38 does not apply under the Approximate Quantities Form.

[5] Local Authorities Editions only.

Issued December 1980

Notes on this Supplement

The Supplement on pages *5 10* has been drawn up so that it may be used for adapting the Standard Form of Building Contract WITH Quantities (or WITH APPROXIMATE Quantities) as follows:

A The recitals in the Contract Form

To be adapted:

EITHER (1)

by inserting 'both . . . and the division of the Works into Sections for phased completion (hereinafter referred to as 'Sections')' in the first recital as shown opposite; in the APPROXIMATE Quantities variant insert 'both' after the words 'and intended to set out a reasonably accurate forecast of the quantity of'.

OR (2) (WITH Quantities only)

by striking out page 5 of the Form, **detaching** page *3* from the Supplement and **affixing** it to the Form so that it may be used as page 5 of the Form. If the **Private edition** is used **insert** as appropriate the address or registered office of the employer.

B The Articles of Agreement and Annexed Conditions

Detach the centre pages *5 8* of this Supplement, which comprise an addition (article 6) to the Articles of Agreement, and **affix** those pages between pages 8 and 9 of the Form so as to follow article 5 and precede the page for signing/sealing.

C To indicate where the Conditions have been modified by the incorporation of the additional Article and Table of Changes **insert** the Symbol Ⓣ (or some other indication) in the margin beside the modified clauses.

D Appendix

Strike out the Appendix in the Form and in substitution **detach** pages *9 10* of this Supplement and **affix** within the Form.

Practice Note 1 on page *11* contains explanatory notes on the Supplement: these are for guidance only and are not intended to form part of the modifications or of the Adapted Contract in which the modifications are incorporated.

Practice Note 1

1   The sectional Completion Supplement (first issued for the 1963 edition by the Joint Contracts Tribunal from December 1975) provides for the Standard Form of Building Contract WITH Quantities (Local Authorities and Private Editions 1980) to be adapted so as to be suitable for use where the Works are to be completed by phased sections. The adaptation makes it necessary to substitute, or add a reference to a 'Section' or 'Sections' in some of the Conditions of the Standard Form where the expression 'the Works' appears. All the adapting modifications are set out seriatim in the Supplement. A Standard Form Contract so adapted is referred to in the following paragraphs as 'the Adapted Contract'.

Practice

2   The Adapted Contract is intended for use only where tenderers are notified that the Employer requires the Works to be carried out by phased sections of which the Employer will take possession on Practical Completion of each Section. The Adapted Contract cannot be used for contracts where the Works are not, at the tender stage, divided into Sections in the tender documents. Attention is called to clause 18 of the Standard Form entitled 'Partial possession by Employer' which is designed to meet those cases where the Adapted Contract has not been used but the Employer nevertheless, by agreement with the Contractor, wishes to take possession of a part of the Works. Where the Adapted Contract is used clause 18 is suitably modified.

3   It is essential that the tender documents (normally the Contract Drawings and the Bills of Quantities) should identify clearly the Sections which, together, comprise the whole Works. The Sections should then be serially numbered and these Section numbers inserted in the Appendix of the Adapted Contract. In this connection care should be taken, in dealing with any part of the Works which is common to all or several Sections, (such as a boiler-house serving three separate Sections, each comprising a block of flats) to put this part of the Works into a separate Section, and to ascribe to it a Section value and Date for Possession, Date for Completion, rate of liquidated damages for delay and Defects Liability Period.

The Adaptation

4   The principal modifications to the contract procedures which are connected with the Adapted Contract are the division of the Works into definite Sections in the Contract Documents, coupled with the division of the Contract Sum into corresponding Section values and the fixing of separate completion periods for each of the Sections. The Adapted Contract remains, however, a single Contract and one Final Certificate is to be issued at the end; no provision is made for separate final certificates for each Section.

5   The contractual provisions of the Standard Form are, therefore, modified in the Adapted Contract as follows:

The Contract Documents

5·1   The first recital to the Articles of Agreement declares that the Contract Drawings and Contract Bills show and describe the division of the Works into Sections. 'Section' is then defined in clause 1·3 by reference to the Sections described by or referred to in the Contract Bills and the Appendix. Clause 2·1 refers to the obligation of the Contractor 'to carry out and complete the Works by Sections'.

5·2   The Section value ascribed to each Section is also entered in the Appendix for the purposes of clauses 18 and 22A. The term 'Section value' is defined in clause 18·1·6 as 'the value ascribed to the relevant Section in the Appendix', and the relevant Appendix entry indicates that each Section value is to be the total value of the Section ascertained from the Contract

Bills. The Section values must amount in total to the Contract Sum, and should take into account the apportionment of Preliminaries and other like items priced in the Contract Bills.

5·3   Also to be entered in the Appendix separately for each Section are: Dates for Possession, Dates for Completion, rates of liquidated damages for delay, and Defects Liability Periods.

Carrying out the Works

5·4   The Contractor is to be given possession of the site for each of the Sections of the Works. He is then to begin and proceed with each Section concurrently or successively as required by and under the Contract, and to carry out each Section within the contract period stated in the Appendix (clause 23·1) or as extended by the Architect[·] under the Contract (clause 25). Liquidated damages for delay relate to each Section (clause 24), and are separately calculated where there is delay in completing any Section.

Practical Completion

5·5   On practical completion of any Section the Architect must issue a practical completion certificate for that Section (clause 17). In consequence the Contractor is relieved of his corresponding duty to insure for that Section under clause 22A (if applicable), and approximately one half of Retention attributable to that Section is released by the operation of clause 30·4. A separate Defects Liability Period operates for each Section (clause 17·2).

Final Account

5·6   When all Sections have been carried out the Architect must issue a certificate to the effect (clause 17·6) and the period begins to run within which the final account is to be prepared and the Final Certificate issued for the whole of the Works comprising all the Sections (clause 30·6).

Insurance

6   The Architect should particularly note that when making insurance arrangements under clause 21·2 agreement must be reached with the insurers as to whether any Section for which a Practical Completion Certificate has been issued is to be treated as continuing to be included in the Works and so not insured OR is to be treated as 'property other than the Works' and so covered by the insurance.

APPROXIMATE Quantities Standard Form

7   The adaptation shown in the Sectional Completion Supplement may also be used in conjunction with the Standard Form of Building Contract for use with Bills of APPROXIMATE Quantities, but varied as necessary where there are differences between that Form and the WITH Quantities Form as indicated in the Supplement.

Footnote   [·] In the Local Authorities edition the term 'Supervising Officer' is substituted throughout where the official concerned is not an Architect.

Explanatory notes

Articles of Agreement	A 50p stamp should be affixed at the top of Page 3 if the Contract is under seal.

Appendix

Fourth Recital and Clause 31	The status of the Employer should be ascertained by striking out an alternative.
Clause 37	Only one clause should remain.
Clause 38.7 or 39.8	Only one clause should apply.
Clause 40.1.1.1	Will be filled in only if Clause 40 is operative.

Note that certain clauses in the Appendix have three columns each with a section number and each of the clauses will be applicable to every section. It is possible that the works will be in more than three sections, in which case a separate page will be attached to the Appendix for the relevant clauses to be separated into the total number of sections.

Scotland

It is not necessary in Scotland to sign a formal document to execute a building contract and until recent years contracts have been let on the basis of a tender and a letter of acceptance. However, both the building contract and the building sub-contracts are prepared on the basis that formal execution of the contract will take place.

If contracts or sub-contracts are let by tender and letter of acceptance in which the terms of the Conditions of the Standard Forms of Building Contract or Sub-contract only are incorporated by reference as they stand, then, most probably English law will be held to apply.

It is therefore in the Contractor's best interests that all contracts should be formal and thus ensure that Scottish law will apply.

Parties to a building contract or sub-contract may well be advised to take legal advice on the method of signing and completion of the Attestation Clause; if this is not suitable the following points should be noted, all of which apply to the Building Contract and Sub-contracts NSC/4/Scot and NSC/4a/Scot.

1   Individuals or partnerships should sign at the foot of the pages indicated, before two witnesses, each of whom signs in the place provided to the left of the signatory's signature, adding the word 'Witness' after their signatures. Their addresses and occupations should also be filled in.
2   Limited companies should execute the contract in accordance with their Memorandum and Articles of Association.
3   If the words 'Both parties consent to registration hereof for preservation and execution' are deleted, then the contract is not liable for stamp duty. Whether or not these words are retained is a matter on which the parties should have legal advice.

Scottish building contract with Scottish Supplement 1980

This grey-coloured form is the formal contract between the Employer and the contractor, and the sample of it which is incorporated here has been filled up by an employer with the necessary deletions and amendments to suit a contract let under the Private Edition with Quantities.

It is essential that the Contractor should carefully check the form and pay particular attention to any deletions and changes made and that it is in accordance with the Appendix II in the Bill of Quantities. Appendix II is normally to be found in the Preliminaries Bill.

Immediately following the sample form are notes on the pages where changes or deletions have taken place.

The numbering of clauses refers to the Standard Conditions of Contract which are dealt with in detail later.

It should also be noted that the Scottish Supplement changing the English version of the Standard Conditions of contract to suit Scottish law is to be found only in this form.

Changes in the April 1982 Revision:

1. Building Contract

Clause 4.1.4	*include*	Clause 4.1
Clause 4.3	*include*	Clause 38.4.3

2. Appendix No. I

Clause 1.4	*add*	definition of specification
Clause 30.11	*delete*	ref. to 30.9.2, *substitute* 30.9.1.2
Clause 35.17	new Clause	
Clause 36.4.8	*add*	ref. to Clause 41
Clause 41.1.1	*delete*	ref. to Clause 3 of Building Contract
Clause 41.1.4	*include*	Clause 4.1
Clause 41.3	*include*	Clause 38.4.3

Scottish Building Contract
(revised April 1982)

SBCC

BUILDING CONTRACT

between

*The Capital Fund Building Society
Capital House, 16-20 Sterling Place
Edinburgh*

and

*Construction Services Ltd
5 Contour Road
Cooper Industrial Estate
Glasgow*

with

SCOTTISH SUPPLEMENT 1980

to

The Conditions of the Standard Form of
Building Contract 1980 Edition

———

The constituent bodies of the
Scottish Building Contract Committee are:

Royal Incorporation of Architects in Scotland
Scottish Building Employers Federation
Scottish Branch of the Royal Institution of Chartered Surveyors
Convention of Scottish Local Authorities
Federation of Specialists and Sub-Contractors
    (Scottish Board)
Committee of Associations of Specialist Engineering
    Contractors (Scottish Branch)
Association of Consulting Engineers (Scottish Group)
Confederation of British Industry
Association of Scottish Chambers of Commerce

———

Copyright of the S.B.C.C., 39 Castle Street, Edinburgh

April 1982

BUILDING CONTRACT

between

*The Capital Fund Building Society
Capital House, 16-20 Sterling Place
Edinburgh*

(hereinafter referred to as 'the Employer)

and

*Construction Services Ltd
5 Contour Road
Cooper Industrial Estate
Glasgow*

(hereinafter referred to as 'the Contractor')

WHEREAS the Employer is desirous of *the erection of a new four-storey office block at 112 Princes Street, Edinburgh, together with all services and external works complete.*

(hereinafter referred to as 'the Works') and the Contractor has offered to carry out and complete the Works for the sum of *One million, two hundred thousand and ten pounds - 39* (£ *1,200,010.39*)

(hereinafter and in the Appendices hereto referred to as 'the Contract Sum') which offer has been or is hereby accepted by the Employer THEREFORE the Employer and the Contractor HAVE AGREED and DO HEREBY AGREE as follows:

1 The Contractor shall carry out the Works in accordance with the Drawings numbered

and the Bills of Quantities ̶/̶Specification annexed and signed as relative hereto.

2 The Works shall be completed in accordance with and the rights and duties of the Employer and the Contractor shall be regulated by

2.1 The Conditions of the Standard Form of Building Contract ̶L̶o̶c̶a̶l̶ ̶A̶u̶t̶h̶o̶r̶i̶t̶i̶e̶s̶*̶/Private Edition with ̶/̶without quantities (1980 Edition) and the supplemental provisions known as the VAT Agreement thereto issued by the Joint Contracts Tribunal which are held to be incorporated in and form part of this Contract, as amended and modified by the provisions contained in the Scottish Supplement forming Appendix No. I hereto; and

2.2 the Abstract of the said Conditions forming Appendix No. II hereto.

3 The term 'the Architect ̶/̶Supervising Officer' shall mean

*Tripartite Design Group
60 West Hanover Street, Glasgow*

and the term 'the Quantity Surveyor' shall mean

*Williams, Smith & Williams
Chartered Quantity Surveyors
17 Portland Close, Glasgow*

and in the event of the Architect/ ̶S̶u̶p̶e̶r̶v̶i̶s̶i̶n̶g̶ ̶O̶f̶f̶i̶c̶e̶r or Quantity Surveyor ceasing to be employed for the purposes of the Contract, the Employer shall nominate another person or

*Delete as required

2

*Delete if Employees of a
Local Authority

persons to the vacant appointment (provided that the Architect~~/Supervising Officer~~ or the *Quantity Surveyor shall not be a person or persons to whom the Contractor shall object for reasons considered to be sufficient by an Arbiter appointed as hereinafter provided): Provided further that no person or persons subsequently appointed to be the Architect~~/Supervising Officer~~ under this Contract shall be entitled to disregard or over-rule any certificate or opinion or decision or approval or instruction given or expressed by the Architect~~/Supervising Officer~~ as the case may be for the time being.

4   In the event of any dispute or difference between the Employer and the Contractor arising during the progress of the Works or after completion or abandonment thereof in regard to any matter or thing whatsoever arising out of this Contract or in connection herewith (but excluding any such dispute or difference arising under Clauses 19A and 31 (to the extent provided in Clause 31.9 and under Clause 3 of the VAT Agreement) the said dispute or difference shall be and is hereby referred to the arbitration of such person as the parties may agree to appoint as Arbiter or failing agreement within 14 days after either party has given to the other written notice to concur in the appointment of an Arbiter as may be appointed by the Sheriff of any Sheriffdom in which the Works or any part thereof are situated: Arbitration proceedings shall be deemed to have been instituted on the date on which the said written notice has been given.

4.1   No arbitration shall commence without the written consent of the parties until after determination or alleged determination of the Contractor's employment or until after Practical Completion or alleged Practical Completion or abandonment of the Works unless it relates to

4.1.1   Clause 3 of this Building Contract

4.1.2   whether or not the issue of an instruction is empowered by the said Conditions

4.1.3   whether or not a certificate has been improperly withheld or is not in accordance with the said Conditions

4.1.4   Clauses 4.1, 25, 32 and 33.

4.2   If the dispute or difference is substantially the same as or is connected with a dispute or difference between

4.2.1   the Employer and a Nominated Sub-Contractor under Agreement NSC/2/Scot or NSC/2a/Scot, or

4.2.2.   the Contractor and any Sub-Contractor,

the Employer and Contractor hereby agree that such dispute or difference shall be referred to an Arbiter appointed or to be appointed to determine the related dispute or difference: Provided that either party may require the appointment of a different Arbiter if he reasonably considers the Arbiter appointed in the related dispute is not suitably qualified to determine the dispute or difference under this Contract.

4.3   Subject to the provisions of Clauses 4.2, 30.9, 38.4.3, 39.5.3 and 40.5 the Arbiter shall have power to

4.3.1   direct such measurements and/or valuations as may in his opinion be desirable in order to determine the rights of the parties

4.3.2   ascertain and amend any sum which ought to have been referred to or included in any certificate

4.3.3   open up review and revise any certificate opinion decision requirement or notice

4.3.4   determine all matters in dispute which shall be submitted to him in the same manner as if no such certificate opinion decision requirement or notice had been given

4.3.5   award compensation or damages and expenses to or against any of the parties to the arbitration.

4.4   The Law of Scotland shall apply to all arbitrations in terms of this clause and the award of the Arbiter shall be final and binding on the parties subject to the provisions of Section 3 of the Administration of Justice (Scotland) Act 1972.

4.5   The Arbiter shall be entitled to remuneration and reimbursement of his outlays.

3

5   This Contract shall be regarded as a Scottish Contract and shall be construed and the rights of parties and all matters arising hereunder determined in all respects according to the Law of Scotland.

6   ~~Both parties consent to registration hereof for preservation and execution~~

IN WITNESS WHEREOF these presents are executed as follows:

Signed by the above named Employer

on the _Sixth_   day of _June_   19 _82_ before these

witnesses

E. Smithers witness ___witness

16 Bramway Road ___address

Barnton

Edinburgh

Office Manager ___occupation

J.F. Cuthbert witness ___witness   _J. Bartholomew_ ___Employer.
(Attention is drawn to the note at the foot of this page)

12A New River Way ___address

S. Queensferry

Fife

Assistant Accountant ___occupation

Signed by the above named Contractor

on the _Sixth_   day of _June_   19 _82_ before these

witnesses

E. Clark witness ___witness

14 Park Street ___address

E. Kilbride

Wages clerk ___occupation

H. Gerald witness ___witness   _J. Palmer_ ___Contractor.
(Attention is drawn to the note at the foot of this page)

4 Hilpark Rd ___address

Bearsden

Glasgow

Surveyor ___occupation

N.B. – This document is set out as for execution by individuals or firms: Where Limited Companies or Local Authorities are involved amendment will be necessary and the appropriate officials should be consulted.

Both parties sign here and on pages 7 and 8.

4

SCOTTISH SUPPLEMENT

(The following are the amendments and modifications to the
Conditions of the Standard Form of Building Contract.
The numbers refer to clauses in the Standard Form).

PART I – GENERAL

1 Interpretation, Definitions, etc.

1.1 and 1.2 shall be deleted.

1.3 The meanings given to the undernoted words and phrases shall be deleted and the following substituted therefor:

Appendix	Appendix No. II to the Building Contract
Arbitrator	Arbiter
Articles or Articles of Agreement	The foregoing Building Contract
Contract Bills	The Bills of Quantities referred to in the Building Contract which have been priced by the Contractor and signed by him and the Employer or on their behalf
Contract Drawings	The drawings referred to in the Building Contract which have been signed by the Employer and the Contractor or on their behalf
Contract Sum	The sum stated in the Building Contract or such other sum as becomes payable in accordance with the Conditions subject to Clause 15.2

Nominated Sub-Contract Documents

Tender NSC/1	Tender NSC/1/Scot
Agreement NSC/2	Agreement NSC/2/Scot
Agreement NSC/2a	Agreement NSC/2a/Scot
Nomination NSC/3	Nomination NSC/3/Scot
Sub-Contract NSC/4	Sub-Contract NSC/4/Scot
Sub-Contract NSC/4a	Sub-Contract NSC/4a/Scot
Works	The works described in the Building Contract and shown and described on Contract Drawings and in Contract Bills

The following clause shall be added:

1.4 Additional definitions:

Execution of this Contract (5.2 and 5.3)	Formal adoption and signing of the Building Contract
Execution of a binding Sub-Contract Agreement (19.3)	Creation of a Sub-Contract
Article 3A, 3B and 4	Clause 3 of the Building Contract
Article 5	Clause 4 of the Building Contract
Real or personal	Heritable or moveable
Section 117 Local Government Act 1972	Section 68 Local Government (Scotland) Act 1973
Specification (Without Quantities Editions only)	The Specification referred to in the Building Contract and signed by Employer and Contractor or on their behalf.

5 Contract Documents – other documents

*Delete as required

*5.3.1.2 shall apply

*5.3.1.2 shall not apply and in Clause 5.3.2 the words '(nor in the master programme for the execution of the Works or any amendment to that programme or revision therein referred to in Clause 5.3.1.2)' shall be deleted.

14   Contract Sum

14.2 line 2 There shall be added after the word 'Conditions,' the words 'including without prejudice thereto Clause 30.11.'

16   Materials and goods unfixed or off-site

16.1 There shall be added at the end 'and for any materials and/or goods purchased prior to their delivery to the site under the separate Contract referred to in Clause 30.3 hereof.'

16.2 shall be deleted.

N.B. – See Clause 30 – Certificates and Payments below.

*Delete as required

22   Insurance of Works against Clause 22 Perils

~~*22A shall apply~~

*22B shall apply.

~~*22C shall apply†~~

N.B.– Clause 22A is applicable to the erection of a new building if the Contractor is required to insure against Clause 22 Perils. Clause 22B is applicable to the erection of a new building if the Employer is to bear the sole risk in respect of Clause 22 Perils. Clause 22C is applicable to alterations of or extensions to an existing building. Therefore strike out Clauses 22B and 22C or Clauses 22A and 22C or Clauses 22A and 22B as may be required.

27   Determination by Employer

27.2 shall be deleted and the following substituted:

27.2.1   In the event of a provisional liquidator being appointed to control the affairs of the Contractor the Employer may determine the employment of the Contractor under this Contract by giving him seven days written notice sent by registered post or recorded delivery of such determination.

27.2.2   In the event of the Contractor becoming bankrupt or making a composition or arrangement with his creditors or having his estate sequestrated or being rendered notour bankrupt or entering into a trust deed for his creditors or having a winding up order made or except for the purposes of reconstruction a resolution for voluntary winding up passed or a receiver or manager of his business or undertaking duly appointed or possession taken by or on behalf of the holder of any debenture secured by a floating charge, the employment of the Contractor under this Contract shall be forthwith automatically determined.

27.2.3   In the event of the employment of the Contractor being determined under Clauses 27.2.1 or 27.2.2 hereof the said employment may be reinstated and continued if the Employer and the Contractor his trustee in bankruptcy, provisional liquidator, liquidator, receiver or manager as the case may be shall so agree.

27.4.3   The words 'and sell any such property of the Contractor' shall be deleted and the words 'and sell any such property so far as belonging to the Contractor' substituted.

28   Determination by Contractor

Clause 28.1.4 (Private Editions only) There shall be added after the word 'creditors' on the first line the words 'or shall have his estate sequestrated or be rendered notour bankrupt or shall enter into a trust deed for his creditors.'

30   Certificates and Payments

30.2.1.3 shall be deleted.

30.3 shall be deleted and the following substituted:

If the Architect ~~Supervising Officer~~ is of the opinion that it is expedient to do so the Employer may enter into a separate contract for the purchase from the Contractor or any Sub-Contractor of any materials and/or goods prior to their delivery to the site, which the Contractor is under obligation to supply in terms of this Contract, and upon such contracts being entered into the purchase of the said materials and/or goods shall be excluded altogether from this Contract and the Contract Sum shall be adjusted accordingly:

Provided that when the Employer enters into a separate contract with any Sub-Contractor

30.3.1   he shall do so only with the consent of the Contractor, which consent shall not be unreasonably withheld, and

30.3.2   payment by the Employer to the Sub-Contractor for any of the said materials and/or goods shall in no way affect any cash discount or other emolument to which the Contractor may be entitled and which shall be paid by the Employer to the Contractor.

The following clause shall be added:

30.11   Nothing in Clauses 30.6.2 or 30.9.1.2 shall prevent the Employer from deducting or adding liquidate and ascertained damages in accordance with Clause 24 hereof from any sum due by him to the Contractor or by the Contractor to the Employer as the case may be under the Final Certificate.

6

PART II – NOMINATED SUB-CONTRACTORS AND NOMINATED SUPPLIERS

35 Nominated Sub-Contractors

35.3 shall be deleted and the following substituted:

The following documents relating to nominated sub-contractors are issued by the Scottish Building Contract Committee and are referred to in the Conditions and in the documents themselves as

Name	Identification No.
Standard Form of Nominated Sub-Contract Tender for use in Scotland	Tender NSC/1/Scot
Agreement between Employer and Nominated Sub-Contractor	Agreement NSC/2/Scot
The above Agreement adapted for use when NSC/1/Scot has not been used	Agreement NSC/2a/Scot
Standard Form for Nomination of Sub-Contractors for use in Scotland	Nomination NSC/3/Scot
Building Sub-Contract for use in Scotland when NSC/1/Scot has been used	Sub-Contract NSC/4/Scot
Building Sub-Contract for use in Scotland when NSC/1/Scot has not been used	Sub-Contract NSC/4a/Scot

and throughout Part 2 of the Conditions, NSC/1/Scot to NSC/3/Scot shall be substituted for NSC/1 to NSC/3 respectively: NSC/4/Scot and NSC/4a/Scot shall be read in conjunction with NSC/4 and NSC/4a respectively.

35.13.5.4.4 shall be deleted and the following substituted:

Clause 35.13.5.3 shall not apply if at the date when the reduction and payment to the Nominated Sub-Contractor referred to in Clause 35.13.5.3 would otherwise be made the Contractor has become bankrupt or made a composition or arrangement with his creditors or had his estate sequestrated or been rendered notour bankrupt or entered into a trust deed for his creditors or had a winding up order made or a resolution for winding up passed (except for the purpose of reconstruction).

35.17 The words 'clause 5' in line 1 and 'clause 4' in line 2 shall be deleted and the words 'clause 8' and 'clause 7' respectively substituted.

35.24.2 shall be deleted and the following substituted:

the Nominated Sub-Contractor becomes bankrupt or makes a composition or arrangement with his creditors or has his estates sequestrated or is rendered notour bankrupt or enters into a Trust Deed for his creditors or has a winding up order made or (except for the purpose of reconstruction) has a resolution for winding up passed or a Receiver or Manager of his business or undertaking appointed or possession taken by or on behalf of any debenture secured by a floating charge; or

The following clause shall be added:

35.27 Determination of employment of Contractor.

The Nominated Sub-Contractor shall recognise an Assignation by the Contractor in favour of the Employer in terms of Clause 27.4.2.1.

36 Nominated Suppliers

36.4.8 shall be deleted and the following substituted:

that if any dispute or difference between the Contractor and the Nominated Supplier is substantially the same as a matter which is in dispute under this Contract then such dispute or difference shall be referred to an Arbiter appointed or to be appointed in terms of Clause 4 of the foregoing Building Contract or under Clause 41 hereof as the case may be and the award of such Arbiter shall be final and binding on the parties.

The following clause shall be added:

36.6 Determination of employment of Contractor.

The Nominated Supplier shall recognise an Assignation by the Contractor in favour of the Employer in terms of Clause 27.4.2.1.

7

PART III – FLUCTUATIONS

(No amendments or modifications are required to this part).

PART IV – ARBITRATIONS

(This part has been added for convenience so that it may be included by reference in the Contract Document when the formal Scottish Building Contract is not being executed).

41 In the event of any dispute or difference between the Employer and the Contractor arising during the progress of the Works or after completion or abandonment thereof in regard to any matter or thing whatsoever arising out of this Contract or in connection herewith (but excluding any such dispute or difference arising under Clauses 19A and 31 (to the extent provided in Clause 31.9 and under Clause 3 of the VAT Agreement) the said dispute or difference shall be and is hereby referred to the arbitration of such person as the parties may agree to appoint as Arbiter or failing agreement within 14 days after either party has given to the other written notice to concur in the appointment of an Arbiter as may be appointed by the Sheriff of any Sheriffdom in which the Works or any part thereof are situated: Arbitration proceedings shall be deemed to have been instituted on the date on which the said written notice has been given.

    41.1 No arbitration shall commence without the written consent of the parties until after determination or alleged determination of the Contractor's employment or until after Practical Completion or alleged Practical Completion or abandonment of the Works unless it relates to

        41.1.1 the nominations of an Architect/~~Supervising Officer~~ or Quantity Surveyor to a vacant appointment

        41.1.2 whether or not the issue of an instruction is empowered by the said Conditions

        41.1.3 whether or not a certificate has been improperly withheld or is not in accordance with the said Conditions

        41.1.4 Clauses 4.1, 25, 32 and 33.

    41.2 If the dispute or difference is substantially the same as or is connected with a dispute or difference between

        41.2.1 the Employer and a Nominated Sub-Contractor under Agreement NSC/2/Scot or NSC/2a/Scot, or

        41.2.2 the Contractor and any Sub-Contractor,

    the Employer and Contractor hereby agree that such dispute or difference shall be referred to an Arbiter appointed or to be appointed to determine the related dispute or difference: Provided that either party may require the appointment of a different Arbiter if he reasonably considers the Arbiter in the related dispute is not suitably qualified to determine the dispute or difference under this Contract.

    41.3 Subject to the provisions of Clauses 4.2, 30.9, 38.4.3, 39.5.3 and 40.5 the Arbiter shall have power to

        41.3.1 direct such measurements and/or valuations as may in his opinion be desirable in order to determine the rights of the parties

        41.3.2 ascertain and amend any sum which ought to have been referred to or included in any certificate

        41.3.3 open up review and revise any certificate opinion decision requirement or notice

        41.3.4 determine all matters in dispute which shall be submitted to him in the same manner as if no such certificate opinion decision requirement or notice had been given

        41.3.5 award compensation or damages and expenses to or against any of the parties to the arbitration.

    41.4 The Law of Scotland shall apply to all arbitrations in terms of this clause and the award of the Arbiter shall be final and binding on the parties subject to the provisions of ~~Clause~~ Section 3 of the Administration of Justice (Scotland) Act 1972.

    41.5 The Arbiter shall be entitled to remuneration and reimbursement of his outlays.

42 This Contract shall be regarded as a Scottish Contract and shall be construed and the rights of parties and all matters arising hereunder determined in all respects according to the Law of Scotland.

_____ Employer   _____ Contractor

8

ABSTRACT OF CONDITIONS

		Clause	
ı Delete as required	Statutory tax deduction scheme – Finance (No. 2) Act 1975	31	Employer at Date of Tender ~~if a 'contractor'~~ is not a 'contractor' for the purposes of the Act and the Regulations

Defects Liability Period (if none other stated is 6 months from the day named in the Certificate of Practical Completion of the Works)   17.2   _12 MONTHS_

Insurance cover for any one occurrence or series of occurrences arising out of one event   21.1.1   £ _5 000 000_

2 No percentage should be inserted if those concerned are all Employees of a Local Authority, but the sum assured should cover the cost of professional services

Percentage to cover Professional fees[2]   22A   _N/A_

Date of Possession   23.1   _4th August 1982_

Date for Completion   1.3   _5th August 1984_

Liquidate and Ascertained Damages   24.2   at the rate of £ _3000_ per _week_

Period of delay:

3 In Clause 28.1.3.2 it is suggested the period should be 3 months: in all other cases 1 month

     by reason of loss or damage caused by any one of the contingencies referred to in Clause 22 Perils (if applicable)[3]   28.1.3.2   _3 MONTHS_

     for any other reason[3]   28.1.3.1, 28.1.3.3 to .7   _1 MONTH_

Period of Interim Certificates (if none stated is one month)   30.1.3   _1 MONTH_

Retention Percentage (if less than five per cent)   30.4.1.1   _3%_

Period of Final Measurement and Valuation (if none stated is six months from the day named in the Certificate of Practical Completion of the Works)   30.6.1.2   _12 MONTHS_

4 Local Authority Editions only

Period for issue of Final Certificate[4] (if none stated is three months)   30.8   _—_

Work reserved for nominated sub-contractors for which the Contractor desires to tender   35.2   _JOINERY FITTINGS_

5 Delete as required

Fluctuations (if alternative required is not shown Clause 38 shall apply)   37   ~~Clause 38~~[5]/ Clause 39[5] /~~Clause 40~~

Percentage addition   ~~38.7 or~~ 39.8   _5_ %

Formula Rules   Rule 3   40.1.1.1   Base Month _—_ 19_—_

6 Not to exceed 10%

     Rule 3   [6]Non-Adjustable Element _—_ % (Local Authority with quantities only)

7 Delete as required

     Rules 10 and 30(i)   Part I[7]/ Part II of Section 2 of the Formula Rules is to apply.

J. Savithdomei _____ Employer.   _J. Mclean_ _____ Contractor.

Explanatory notes

Clause 1   Here either Bills of Quantities or Specification should have been deleted.

Clause 2   Here the alternative forms of the Conditions are deleted leaving the Standard Form which is operative for this contract.

Clause 3   There are four references where either Architect or Supervising Officer should be deleted.

Clause 6   This is the Attestation where the words at the beginning of the clause are either left in or deleted. If left in then the Contract is liable for stamp duty. Again it is suggested that legal advice should be sought on this matter.

Scottish Supplement

Clause 5   Here a deletion should be made to establish whether or not Clause 5.3.1.2 is operative. This refers to the Contractor's Master Programme.

Clause 22   Only one of the three alternatives 22A, 22B or 22C should be left in and it should be noted at whose risk the insurance has to be effected.

Clause 30   There is a reference here to Architect or Supervising Officer.

Appendix II

Clause 31   Here the status of the Employer is ascertained.

Clause 37   Two alternatives to the fluctuation clauses should be deleted.
The percentage addition should refer to the appropriate clause.

Scottish Building Contract with Scottish Supplement for Sectional Completion 1980

This buff-coloured form is the formal contract between the Employer and the Contractor where sectional completion has been agreed at the tender stage.

A sample form is incorporated here in 12 pages which has been filled up by an employer with the necessary deletions and amendments to suit a contract let under the Private Edition with Quantities.

Again it is very important for the Contractor to check the form with particular attention to any deletions and changes made and that it accords with the Appendix II in the Preliminaries Bill. Appendix II is this time under the heading of Abstract of Conditions, split into columns to represent the number of sections under which the works have to be carried out, and these sections are numbered or otherwise identified.

Following are notes on the pages where changes or deletions have taken place and immediately after the sample form are guidance notes issued by the Scottish Building Contract Committee. Again, the numbering of the clauses refers to the Standard Conditions of Contract.

In this version of the Contract the Scottish Supplement not only caters for the changes required to accord with the law of Scotland, but also changes the wording of the relevant clauses to accord with executing the works in sections.

Apart from the Bill of Quantities where the layout of the Bill will be in the various sections of the work, the Scottish Supplement for Sectional Completion is the only document where the changes are recorded to the relevant clauses and is so incorporated into the Contract.

Note that Clause 18 – Partial possession by Employer, is not intended as an alternative to the Sectional Completion Supplement. Where this form is used, the sections have been pre-determined at tender stage. Clause 18 is used where possession of part of the works was not planned at tender stage and was not an obligation imposed by the Employer.

N.B.

April 1981 Revision

As compared with the original the following alterations have been made.

Building Contract
    Clause 4.1.4 – Add reference to 4.1.
    Clause 4.3 – Add reference to 38.4.3.

Appendix No. I
    2 – Contractor's obligations:
        Amendment to Clause 2.1.
    30 – Certificate and payments
        30.11 Delete 30.9.2 substitute 30.9.1.2.
    35 – Nominated Sub-Contractor
        Add: amendment to 35.17.
    38 – Contribution, levy, tax fluctuations
        Add amendments.
    39 – Labour, materials, cost and tax fluctuations
        Add amendments.
    41 – Arbitrations
        41.1.1 Re-drafted.
        41.1.4 Add reference to 4.1.
        41.3 Add reference to 38.4.3.

Scottish Building Contract
Sectional Completion Supplement
(Revised April 1981)

SBCC

BUILDING CONTRACT

between

The Capital Fund Building Society
Capital House
16-20 Sterling Place
Edinburgh

and

Construction Services Ltd
5 Contour Road
Cooper Industrial Estate
Glasgow

with

SCOTTISH SUPPLEMENT
FOR
SECTIONAL COMPLETION
1980

to

The Conditions of the Standard Form of
Building Contract 1980 Edition
(with quantities only)

The constituent bodies of the
Scottish Building Contract Committee are:

Royal Incorporation of Architects in Scotland
Scottish Building Employers Federation
Scottish Branch of the Royal Institution of Chartered Surveyors
Convention of Scottish Local Authorities
Federation of Specialists and Sub-Contractors
   (Scottish Board)
Committee of Associations of Specialist Engineering
   Contractors (Scottish Branch)
Association of Consulting Engineers (Scottish Group)
Confederation of British Industry
Association of Scottish Chambers of Commerce

Copyright of the S.B.C.C., 39 Castle Street, Edinburgh

April 1981

Scottish Building Contract
Sectional Completion Edition
(Revised April 1981)

BUILDING CONTRACT

between

The Capital-Fund Building Society
Capital House
16-20 Sterling Place
Edinburgh

(hereinafter referred to as 'the Employer')

and

Construction Services Ltd
5 Contour Road
Cooper Industrial Estate
Glasgow

(hereinafter referred to as 'the Contractor')

WHEREAS the Employer is desirous of *alterations to form a four-storey office block at 112 Princes St, Edinburgh, from two adjacent buildings complete with link connections*

(hereinafter referred to as 'the Works') and the Contractor has offered to carry out and complete the Works for the sum of *One million, five hundred and twenty thousand, three hundred and sixteen pounds - 18* (£ *527 316-18*)

(hereinafter and in the Appendices hereto referred to as 'the Contract Sum') which offer has been or is hereby accepted by the Employer THEREFORE the Employer and the Contractor HAVE AGREED and DO HEREBY AGREE as follows:

1   The Contractor shall carry out the Works in accordance with the Drawings numbered

and the Bills of Quantities ~~/Specification~~ annexed and signed as relative hereto.

2   The Works shall be completed in accordance with and the rights and duties of the Employer and the Contractor shall be regulated by

2.1   The Conditions of the Standard Form of Building Contract ~~Local Authorities*/~~Private Edition with quantities (1980 Edition) and the supplemental provisions known as the VAT Agreement thereto issued by the Joint Contracts Tribunal which are held to be incorporated in and form part of this Contract, as amended and modified by the provisions contained in the Scottish Supplement for Sectional Completion forming Appendix No. I hereto; and

2.2   the Abstract of the said Conditions forming Appendix No. II hereto.

3   the term 'the Architect*/~~Supervising Officer~~' shall mean

Tripartite Design Group
60 West Hanover Street
Glasgow

and the term 'the Quantity Surveyor' shall mean

Williams, Smith + Williams
Chartered Quantity Surveyors
17 Portland Close, Glasgow

and in the event of the Architect/~~Supervising Officer~~ or Quantity Surveyor ceasing to be employed for the purposes of the Contract, the Employer shall nominate another person or

*Delete as required

2

*Delete if Employees of a
Local Authority

persons to the vacant appointment (provided that *the Architect/Supervising Officer or the *Quantity Surveyor shall not be a person or persons to whom the Contractor shall object for reasons considered to be sufficient by an Arbiter appointed as hereinafter provided): Provided further that no person or persons subsequently appointed to be the Architect/Supervising Officer under this Contract shall be entitled to disregard or over-rule any certificate or opinion or decision or approval or instruction given or expressed by the Architect/Supervising Officer as the case may be for the time being.

4 In the event of any dispute or difference between the Employer and the Contractor arising during the progress of the Works or after completion or abandonment thereof in regard to any matter or thing whatsoever arising out of this contract or in connection herewith (but excluding any such dispute or difference arising under Clauses 19A and 31 (to the extent provided in Clause 31.9 and under Clause 3 of the VAT Agreement) the said dispute or difference shall be and is hereby referred to the arbitration of such person as the parties may agree to appoint as Arbiter or failing agreement within 14 days after either party has given to the other written notice to concur in the appointment of an Arbiter as may be appointed by the Sheriff of any Sheriffdom in which the Works or any part thereof are situated: Arbitration proceedings shall be deemed to have been instituted on the date on which the said written notice has been given.

4.1 No arbitration shall commence without the written consent of the parties until after determination or alleged determination of the Contractor's employment or until after Practical Completion or alleged Practical Completion of the Works or of any relevant Section or abandonment of the Works unless it relates to

4.1.1 Clause 3 of this Building Contract

4.1.2 whether or not the issue of an instruction is empowered by the said Conditions

4.1.3 whether or not a certificate has been improperly withheld or is not in accordance with the said Conditions

4.1.4 Clauses 4.1, 25, 32 and 33.

4.2 If the dispute or difference is substantially the same as or is connected with a dispute or difference between

4.2.1 the Employer and a Nominated Sub-Contractor under Agreement NSC/2/Scot or NSC/2a/Scot, or

4.2.2 the Contractor and any Sub-Contractor,

the Employer and Contractor hereby agree that such dispute or difference shall be referred to an Arbiter appointed or to be appointed to determine the related dispute or difference: Provided that either party may require the appointment of a different Arbiter if he reasonably considers the Arbiter appointed in the related dispute is not suitably qualified to determine the dispute or difference under this Contract.

4.3 Subject to the provisions of Clauses 4.2, 30.9, 38.4.3, 39.5.3 and 40.5 the Arbiter shall have power to

4.3.1 direct such measurements and/or valuations as may in his opinion be desirable in order to determine the rights of the parties

4.3.2 ascertain and amend any sum which ought to have been referred to or included in any certificate

4.3.3 open up review and revise any certificate opinion decision requirement or notice

4.3.4 determine all matters in dispute which shall be submitted to him in the same manner as if no such certificate opinion decision requirement or notice had been given

4.3.5 award compensation or damages and expenses to or against any of the parties to the arbitration.

4.4 The Law of Scotland shall apply to all arbitrations in terms of this clause and the award of the Arbiter shall be final and binding on the parties subject to the provisions of Section 3 of the Administration of Justice (Scotland) Act 1972.

4.5 The Arbiter shall be entitled to remuneration and reimbursement of his outlays.

3

5   This Contract shall be regarded as a Scottish Contract and shall be construed and the rights of parties and all matters arising hereunder determined in all respects according to the Law of Scotland.

6   Both parties consent to registration hereof for preservation and execution:

IN WITNESS WHEREOF these presents are executed at

on the _Sixth_   day of _June_   19_82_ before these

witnesses subscribing

S Smithers Witness witness
16 Browning Road
Burnton, Edinburgh address
Office manager occupation

J Bartholomew Employer.

T F Cuthbert Witness witness
112A New River Way
S Queensferry, Fife address
Assistant Accountant occupation

E Clark Witness witness
14 Park Street
E Kilbride address
Wages Clerkess occupation

J Milner Contractor.

H Goodall Witness witness
4 Thornpark Road
Bearsden, Glasgow address
Surveyor occupation

N.B. – This document is set out as for execution by individuals or firms: Where Limited Companies or Local Authorities are involved amendment will be necessary and the appropriate officials should be consulted.

Both parties sign here and on pages 9 and 11.

4

SCOTTISH SUPPLEMENT
FOR
SECTIONAL COMPLETION

(The following are the amendments and modifications to the
Conditions of the Standard Form of Building Contract.
The numbers refer to clauses in the Standard Form).

PART 1 – GENERAL

1   Interpretation, Definitions, etc.

1.1 and 1.2 shall be deleted.

1.3   The meanings given to the undernoted words and phrases shall be deleted and the following substituted therefor:

Appendix	Appendix No. II to the Building Contract
Arbitrator	Arbiter
Articles or Articles of Agreement	The foregoing Building Contract
Contract Bills	The Bills of Quantities referred to in the Building Contract which have been priced by the Contractor and signed by him and the Employer or on their behalf
Contract Drawings	The drawings referred to in the Building Contract which have been signed by the Employer and the Contractor or on their behalf
Contract Sum	The sum stated in the Building Contract or such other sum as becomes payable in accordance with the Conditions subject to Clause 15.2

Nominated Sub-Contract Documents

Tender NSC/1	Tender NSC/1/Scot
Agreement NSC/2	Agreement NSC/2/Scot
Agreement NSC/2a	Agreement NSC/2a/Scot
Nomination NSC/3	Nomination NSC/3/Scot
Sub-Contract NSC/4	Sub-Contract NSC/4/Scot
Sub-Contract NSC/4a	Sub-Contract NSC/4a/Scot

Works	The works described in the Building Contract and shown and described on the Contract Drawings and in the Contract Bills

The following clause shall be added:

1.4   Additional definitions:

Completion Date, Date for Completion, Date of Possession Defects Liability Period	The definitions given in Clause 1.3 shall apply for each Section
Execution of this Contract (5.2 and 5.3)	Formal adoption and signing of the Building Contract
Execution of a binding Sub-Contract Agreement (19.3)	Creation of a Sub-Contract
Article 3A, 3B and 4	Clause 3 of the Building Contract
Article 5	Clause 4 of the Building Contract
Real or personal	Heritable or moveable
Section 117 Local Government Act 1972	Section 68 Local Government (Scotland) Act 1973.
Section	One of the Sections into which the Works have been divided for phased completion as shown upon the Contract Drawings and described by or referred to in the Contract Bills and in Appendix No. II.

5

2　Contractor's obligations

2.1　line 1 There shall be added after 'the Works,' the words 'by Sections.'

5　Contract Documents – other documents

*Delete as required

~~*5.3.1.2 shall apply.~~

*5.3.1.2 shall not apply and in Clause 5.3.2 the words '(nor in the master programme for the execution of the Works or any amendment to that programme or revision therein referred to in Clause 5.3.1.2)' shall be deleted.

14　Contract Sum

14.2　line 2 There shall be added after the word 'Conditions,' the words 'including without prejudice thereto Clause 30.11.'

16　Materials and goods unfixed or off-site

16.1　There shall be added at the end 'and for any materials and/or goods purchased prior to their delivery to the site under the separate Contract referred to in Clause 30.3 hereof.'

16.2　shall be deleted.

N.B. – See Clause 30 – Certificates and Payments below.

17　Practical Completion and Defects Liability

17.1　line 1 The words 'the Works' shall be deleted and the words 'any Section' substituted.

line 2 The words 'the Works' shall be deleted and the words 'that Section' substituted.

17.2　line 1 The words 'within the Defects Liability Period' shall be deleted, and the words 'in any Section within the Defects Liability Period in relation thereto' substituted

line 3 There shall be added after 'Completion of' the words 'that Section of'

line 4 There shall be added after 'schedule of defects' the words 'for that Section'

17.3　line 3 The words 'within the Defects Liability Period' shall be deleted and the words 'in any Section within the Defects Liability Period in relation thereto' substituted

line 4 The words 'the Works' shall be deleted and the words 'that Section' substituted

17.4　line 3 There shall be added after 'making good defects' the words 'in the relevant Section'

17.5　lines 2 and 3 There shall be added after 'Practical Completion' the words 'of any Section' and 'of that Section' respectively.

The following clause shall be added:

17.6　When in the opinion of the Architect ~~(Supervising Officer)~~ all the Sections have been practically completed he shall forthwith issue (in addition to any certificates of practical completion of the Sections) a Certificate of Practical Completion of the Works and practical completion of the whole of the Works shall for the purpose of Clause 30.6.1.1 be deemed to have taken place on the day named in such Certificate.

18　Partial possession by Employer

18.1　line 1 The words 'the Works' shall be deleted and the words 'any Section' substituted

18.1.5　line 2 The words 'the Works' shall be deleted and the words 'any Section' substituted
lines 5 and 6 The words 'Contract Sum' shall be deleted and the words 'Section Value' substituted.

The following clause shall be added:

18.1.6　For the purpose of Clause 18 'Section Value' shall mean the value ascribed to the relevant Section in Appendix No. II hereto.

22　Insurance of Works against Clause 22 Perils

*Delete as required

~~(*Clause 22A~~ shall apply amended as follows: Add as 22A.3.3 'Upon Practical Completion of any Section that Section shall be at the sole risk of the Employer as regards any of the Clause 22 Perils and after taking into account any operation of the provisions of Clause 18.1.4 the Contractor shall reduce the value insured under Clause 22A.1 by the value ascribed to that Section in Appendix No. II hereto or such other value as may be specifically agreed in writing and recorded by amendment of the relevant part of said Appendix No. II.'

~~*Clause 22B shall apply.~~

*Clause 22C shall apply.

N.B. – Clause 22A is applicable to the erection of a new building if the Contractor is required to insure against the Clause 22 Perils. Clause 22B is applicable to the erection of a new building if the Employer is to bear the sole risk in respect of the Clause 22 Perils. Clause 22C is applicable to alterations of or extensions to an existing building. Therefore strike out Clauses 22B and 22C or Clauses 22A and 22C or Clauses 22A and 22B as may be required.

6

23 Date of Possession, completion and postponement

23.1 line 1 The words 'possession of the site' shall be deleted, and the words 'in relation to any Section possession of the relevant part of the site' substituted
line 2 The words 'the Works' shall be deleted and the words 'that Section' substituted
line 3 There shall be added after 'Completion Date' the words 'in relation thereto.'

24 Damages for non-completion

24.1 line 1 The words 'the Works' shall be deleted, and the words 'any Section' substituted and There shall be added after 'Completion Date' the words 'in relation thereto.'

24.2.1 line 4 The words 'calculated at the rate stated in the Appendix' shall be deleted, and the words 'calculated at the rate stated in relation thereto in Appendix No. II' substituted
line 5 There shall be added after 'Completion Date' the words 'of that Section'
line 6 There shall be added after 'Practical Completion' the words 'in relation thereto.'

24.2.2 line 1 There shall be added after 'Completion Date' the words 'for any Section.'

25 Extension of time

25.2.1.1 line 1 The words 'the Works' shall be deleted, and the words 'any Section' substituted.

25.2.2.2 line 1 The words 'the Works beyond the Completion Date' shall be deleted and the words 'that Section beyond the Completion Date in relation thereto' substituted.

25.3.1.2 shall be deleted and the following substituted:
'the completion of any Section or Sections is likely to be delayed thereby beyond the Completion Date in relation thereto.'

25.3.2 line 2 There shall be added after 'Completion Date' the words 'for that Section.'
line 5 There shall be added after 'omission' the words 'from that Section.'
line 7 There shall be added after 'extension of time' the words 'for that Section.'

25.3.3 line 1 There shall be added after 'Practical Completion' the words 'of any Section.'

25.3.3.1 line 1 There shall be added after 'Completion Date' the words 'for any Section.'

25.3.3.2 line 1 There shall be added after 'Completion Date' the words 'for any Section.'
line 3 There shall be added after 'omission' the words 'from that Section.'
line 5 There shall be added after 'extension of time' the words 'for that Section.'

25.3.4.1 shall be deleted and the following substituted:
'The Contractor shall use constantly his best endeavours to prevent delay in the progress of any Section, howsoever caused, and to prevent the completion of any Section being delayed or further delayed beyond the Completion Date in relation thereto.'

25.3.4.2 line 2 There shall be added after 'the Works' the words 'or any Section thereof.'

25.3.5 line 3 There shall be added after 'Completion Date' the words 'for any Section.'

25.3.6 shall be deleted and the following substituted:
'No decision of the Architect/Supervising Officer under Clause 25.3 shall fix a Completion Date for any Section earlier than the Completion Date in relation thereto stated in Appendix No. II.'

25.4.6 line 4 There shall be added after 'Completion Date' the words 'for any Section.'

25.4.12 line 1 There shall be added after 'the Works' the words 'or any Section.'

26 Loss and expense caused by matters materially affecting regular progress of the Works

26.1 and 26.1.1 The words 'regular progress of the Works or of any part' shall be deleted and the words 'regular progress of the Works or of any Section or part' substituted.

26.4.2 line 4 The words 'Sub-Contract Works or of any part thereof' shall be deleted and the words 'Sub-Contract Works or of any Section or part thereof' substituted.

27 Determination by Employer

27.2 shall be deleted and the following substituted:

27.2.1 In the event of a provisional liquidator being appointed to control the affairs of the Contractor the Employer may determine the employment of the Contractor under this Contract by giving him seven days written notice sent by registered post or recorded delivery of such determination.

27.2.2 In the event of the Contractor becoming bankrupt or making a composition or arrangement with his creditors or having his estate sequestrated or being rendered notour bankrupt or entering into a trust deed for his creditors or having a winding up order made or except for the purposes of reconstruction a resolution for voluntary winding up passed or a receiver or manager of his business or undertaking duly appointed or possession taken by or on behalf of the holder of any debenture secured by a floating charge, the employment of the Contractor under this Contract shall be forthwith automatically determined.

7

27.2.3 In the event of the employment of the Contractor being determined under Clauses 27.2.1 or 27.2.2 hereof the said employment may be reinstated and continued if the Employer and the Contractor his trustee in bankruptcy, provisional liquidator, liquidator, receiver or manager as the case may be shall so agree.

27.4.3 line 6 The words 'and sell any such property of the Contractor' shall be deleted and the words 'and sell any such property so far as belonging to the Contractor' substituted.

28 Determination by Contractor

28.1.4 (Private Editions only) There shall be added after the word 'creditors' on the first line the words ' or shall have his estate sequestrated or be rendered notour bankrupt or shall enter into a trust deed for his creditors.'

29 Works by Employer or persons employed or engaged by Employer

29.1 line 4 There shall be added after 'complete the Works' the words 'or any Section thereof.'

30 Certificates and Payments

30.1.3 line 2 There shall be added after the words 'Certificate of Practical Completion' the words 'of the whole of the Works under Clause 17.6.'

30.2.1.3 shall be deleted.

30.3 shall be deleted and the following substituted:

If the Architect/Supervising Officer is of the opinion that it is expedient to do so the Employer may enter into a separate contract for the purchase from the Contractor or any Sub-Contractor of any materials and/or goods prior to their delivery to the site, which the Contractor is under obligation to supply in terms of this Contract, and upon such contracts being entered into the purchase of the said materials and/or goods shall be excluded altogether from this Contract and the Contract Sum shall be adjusted accordingly:

Provided that when the Employer enters into a separate contract with any Sub-Contractor

30.3.1 he shall do so only with the consent of the Contractor, which consent shall not be unreasonably withheld, and

30.3.2 payment by the Employer to the Sub-Contractor for any of the said materials and/or goods shall in no way affect any cash discount or other emolument to which the Contractor may be entitled and which shall be paid by the Employer to the Contractor.

30.4 There shall be added at the start of the clause

'The following provisions of Clause 30.4 shall apply severally in respect of each Section.'

30.5 There shall be added after the words 'following rules,' the words 'which shall apply severally to amounts retained in respect of each Section.'

30.6.1.1 line 1 The words 'Practical Completion of the Works' shall be deleted and the words 'the issue by the Architect/Supervising Officer of the Certificate of Practical Completion as required by Clause 17.6 (or where relevant the Certificate of Practical Completion of any Section)' substituted.

30.6.1.2 line 4 There shall be added after the words 'within the Period' the words '(or Periods, if Final Measurement and Valuation is to be completed in respect of each Section).'

30.8 (Local Authorities Editions only) line 1 The words 'the period the length of which is stated in the Appendix' shall be deleted and the words 'three months' substituted.

line 2 There shall be added after the words 'stated in the Appendix' the words 'in respect of the Section last completed.'

(Local Authorities and Private Editions) line 3 There shall be added after the words 'clause 17' the words 'in respect of all Sections.'

The following clause shall be added:

30.11 Nothing in Clauses 30.6.2 or 30.9.1.2 shall prevent the Employer from deducting or adding liquidate and ascertained damages in accordance with Clause 24 hereof from any sum due by him to the Contractor or by the Contractor to the Employer as the case may be under the Final Certificate.

32 Outbreak of hostilities

32.1.2 line 1 There shall be added after the words 'the Works or any' the words 'Section or.'

33 War damage

33.1 line 1 There shall be added after the words 'the Works or any' the words 'Section or.'

33.3 line 3 There shall be added after the words 'the Works or any,' the words 'Section or.'

8

PART 2 – NOMINATED SUB-CONTRACTORS AND NOMINATED SUPPLIERS

35 Nominated Sub-Contractors

35.3 shall be deleted and the following substituted:

The following documents relating to nominated sub-contractors are issued by the Scottish Building Contract Committee and are referred to in the Conditions and in the documents themselves as

Name	Identification No.
Standard Form of Nominated Sub-Contract Tender for use in Scotland	Tender NSC/1/Scot
Agreement between Employer and Nominated Sub-Contractor	Agreement NSC/2/Scot
The above Agreement adapted for use when NSC/1/Scot has not been used	Agreement NSC/2a/Scot
Standard Form for Nomination of Sub-Contractors for use in Scotland	Nomination NSC/3/Scot
Building Sub-Contract for use in Scotland when NSC/1/Scot has been used	Sub-Contract NSC/4/Scot
Building Sub-Contract for use in Scotland when NSC/1/Scot has not been used	Sub-Contract NSC/4a/Scot

and throughout Part 2 of this Appendix, NSC/1/Scot to NSC/3/Scot shall be substituted for NSC/1 to NSC/3 respectively: NSC/4/Scot and NSC/4a/Scot shall be read in conjunction with NSC/4 and NSC/4a respectively.

35.13.5.4.4 shall be deleted and the following substituted:

Clause 35.13.5.3 shall not apply if at the date when the reduction and payment to the Nominated Sub-Contractor referred to in Clause 35.13.5.3 would otherwise be made the Contractor has become bankrupt or made a composition or arrangement with his creditors or had his estate sequestrated or been rendered notour bankrupt or entered into a trust deed for his creditors or had a winding up order made or a resolution for winding up passed (except for the purpose of reconstruction).

35.17 The words 'clause 5' in line 1 and 'clause 4' in line 2 shall be deleted and the words 'clause 8' and 'clause 7' respectively substituted.

35.24.2 shall be deleted and the following substituted:

the Nominated Sub-Contractor becomes bankrupt or makes a composition or arrangement with his creditors or has his estates sequestrated or is rendered notour bankrupt or enters into a Trust Deed for his creditors or has a winding up order made or (except for the purpose of reconstruction) has a resolution for winding up passed or a Receiver or Manager of his business or undertaking appointed or possession taken by or on behalf of any debenture secured by a floating charge; or

The following clause shall be added:

35.27 Determination of employment of Contractor.

The Nominated Sub-Contractor shall recognise an Assignation by the Contractor in favour of the Employer in terms of Clause 27.4.2.

36 Nominated Suppliers

36.4.8 shall be deleted and the following substituted:

that if any dispute or difference between the Contractor and the Nominated Supplier is substantially the same as a matter which is in dispute under this Contract then such dispute or difference shall be referred to an Arbiter appointed or to be appointed in terms of Clause 4 of the foregoing Building Contract or under Clause 41 hereof as the case may be and the award of such Arbiter shall be final and binding on the parties.

The following clause shall be added:

36.6 Determination of employment of Contractor.

The Nominated Supplier shall recognise an Assignation by the Contractor in favour of the Employer in terms of Clause 27.4.2.

PART 3 – FLUCTUATIONS

38 Contribution, levy and tax fluctuations

38.4.7   line 4 There shall be added after '38.3,' the words 'for any Section.'
         line 6 There shall be added after 'Completion Date,' the words 'in relation thereto.'

38.4.8   There shall be added after 'applied' the words 'in relation to any Section.'

38.4.8.1 There shall be added after 'Clause 25,' the words 'as hereinbefore modified.'

38.4.8.2 There shall be added after 'Completion Date,' the words 'for that Section.'

39 Labour and materials cost and tax fluctuations

39.5.7   line 4 There shall be added after '39.4,' the words 'for any Section.'
          line 6 There shall be added after 'Completion Date,' the words 'in relation thereto.'

39.5.8   There shall be added after 'applied,' the words 'in relation to any Section.'

39.5.8.1 There shall be added after 'Clause 25,' the words 'as hereinbefore modified.'

39.5.8.2 There shall be added after 'Completion Date' the words 'for that Section.'

40 Use of Price Adjustment Formulae

40.7.1.1 The words 'If the Contractor fails to complete the Works by the Completion Date, formula adjustment of the Contract Sum under Clause 40' shall be deleted and the words 'If the Contractor fails to complete any Section by the Completion Date in relation thereto formula adjustment of the Contract Sum under Clause 40 relevant to that Section' substituted.

40.7.2.1 There shall be added after the words 'the printed text of Clause 25' the words 'as hereinbefore amended.'

PART 4 – ARBITRATIONS

(The following Part has been added for convenience only so that it may be included by reference in the contract documents when the formal Scottish Building Contract is not being executed).

41   In the event of any dispute or difference between the Employer and the Contractor arising during the progress of the Works or after completion or abandonment thereof in regard to any matter or thing whatsoever arising out of this Contract or in connection herewith (but excluding any such dispute or difference arising under Clauses 19A and 31 (to the extent provided in Clause 31.9 and under Clause 3 of the VAT Agreement) the said dispute or difference shall be and is hereby referred to the arbitration of such person as the parties may agree to appoint as Arbiter or failing agreement within 14 days after either party has given to the other written notice to concur in the appointment of an Arbiter as may be appointed by the Sheriff of any Sheriffdom in which the Works or any part thereof are situated: Arbitration proceedings shall be deemed to have been instituted on the date on which the said written notice has been given.

41.1   No arbitration shall commence without the written consent of the parties until after determination or alleged determination of the Contractor's employment or until after Practical Completion or alleged Practical Completion of the Works or of any relevant Section or abandonment of the Works unless it relates to

41.1.1   the nomination of an Architect/~~Supervising Officer~~ or Quantity Surveyor to a vacant appointment

41.1.2   whether or not the issue of an instruction is empowered by the said Conditions

41.1.3   whether or not a certificate has been improperly withheld or is not in accordance with the said Conditions

41.1.4   Clauses 4.1, 25, 32 and 33.

41.2   If the dispute or difference is substantially the same as or is connected with a dispute or difference between

41.2.1   the Employer and a Nominated Sub-Contractor under Agreement NSC/2/Scot or NSC/2a/Scot, or

41.2.2   the Contractor and any Sub-Contractor,

the Employer and Contractor hereby agree that such dispute or difference shall be referred to an Arbiter appointed or to be appointed to determine the related dispute or difference: Provided that either party may require the appointment of a different Arbiter if he reasonably considers the Arbiter in the related dispute is not suitably qualified to determine the dispute or difference under this Contract.

41.3   Subject to the provisions of Clauses 4.2, 30.9, 38.4.3, 39.5.3 and 40.5 the Arbiter shall have power to

41.3.1   direct such measurements and/or valuations as may in his opinion be desirable in order to determine the rights of the parties

41.3.2   ascertain and amend any sum which ought to have been referred to or included in any certifcate

41.3.3   open up review and revise any certificate opinion decision requirement or notice

41.3.4   determine all matters in dispute which shall be submitted to him in the same manner as if no such certificate opinion decision requirement or notice had been given

41.3.5   award compensation or damages and expenses to or against any of the parties to the arbitration.

41.4   The Law of Scotland shall apply to all arbitrations in terms of this clause and the award of the Arbiter shall be final and binding on the parties subject to the provisions of Section 3 of the Administration of Justice (Scotland) Act 1972.

41.5   The Arbiter shall be entitled to remuneration and reimbursement of his outlays.

42   This Contract shall be regarded as a Scottish Contract and shall be construed and the rights of parties and all matters arising hereunder determined in all respects according to the Law of Scotland.

_____ Employer   _____ Contractor

ABSTRACT OF CONDITIONS

		Clause	
*1 **Delete as required**	Statutory tax deduction scheme – Finance (No. 2) Act 1976	31	Employer at Date of Tender ~~is 'a contractor' *1~~ is not 'a contractor' for purposes of the Act and the Regulations

			*2Section number	Section number	Section number
*2 **If the Works are in more than three Sections a separate sheet should be used**	Section of Works as shown on the Contract Drawings and described in the Contract Bills	2.1	*1*	*2*	*3*
	Section Value (total value of Section ascertained from Contract Bills)	18.1.5, 22A	£13 658·9	£13 658·9	£00 000
	Defects Liability period (if none stated is 6 months from the day named in the Certificate of Practical Completion of Section)	17, 18, 30	12 months	12 months	12 months
	Date of Possession of Section	23.1	4/8/82	4/8/82	4/12/82
	Date for Completion of Section	1.3	5/8/84	16/9/83	16/9/83
	Rate of liquidate and ascertained damages for Section	24.2.1	£ 1450 per WEEK	£ 1450 per WEEK	£ 100 per WEEK
	Insurance cover for any one occurrence or series of occurrences arising out of one event	21.1.1	£5 000 000		
*3 **Where the professional persons concerned with the Works are all employees of a Local Authority no percentage should be inserted, but care should be taken to include in the sum assured the cost to the Employer of their services**	Percentage to cover professional fees *3	22A	N/A %		

*4 **It is suggested that the periods should be (i) three months and (ii) one month. It is essential that periods be inserted since otherwise no period of delay would be prescribed**

Period of Delay*4

(i) by reason of loss or damage caused by any one of the contingencies referred to in Clause 22A or Clause 22B (if applicable) — 28.1.3.2 — 3 months

(ii) for any other reason — 28.1.3.1, 28.1.3.3 to .7 — 1 month

Period of Interim Certificates (if none stated is one month) — 30.1.3 — 1 month

Retention Percentage (if less than 5 per cent) *5 — 30.4.1.1 — 3%

*5 **The percentage will be five per cent unless a lower rate is specified here**

Period of Final Measurement (if none stated is 6 months from the day named in the Certificate of Practical Completion of the Works or of any Section where so agreed) — 30.6.1.2 — 12 months

*6 **Local Authority Editions only**

*6Period for issue of Final Certificate (if none stated is 3 months) — 30.8 — JOINERY FITTINGS

11

	Clause	
Work reserved for Nominated Sub-Contractors for which the Contractor desires to tender	35.2	_____

*7 Delete as required

Fluctuations (if alternative required is not shown Clause 38 shall apply)   ~~Clause 38*7~~/ Clause 39*7 /~~Clause 40~~ shall apply

Percentage addition   ~~38.7 or~~ 39.8 _____ 5 _____ %

Formula Rules   40.1.1.1

Rule 3   Base Month _____ — _____ 19____

Rule 3   Non-adjustable element _____ — _____%
(Local Authority Contract only) *8

*8 Not to exceed 10%

*9 Delete as required

Rules 10 and 30(i)   ~~Part I *9/~~Part II of Section 2 of the Formula Rules is to apply/

_____ Employer   _____ Contractor

THE SCOTTISH BUILDING CONTRACT COMMITTEE

Contracts where Phased Completion by Sections has been agreed at the Tender Stage

NOTES FOR GUIDANCE

The Joint Contracts Tribunal has published a Supplement to the 1980 Standard Form of Building Contract for use when Sectional Completion has been agreed at the Tender stage, and consequently the SBCC has up-dated its existing Sectional Completion Supplement in the same manner. As before the Scottish Building Contract for Sectional Completion combines the amendments required for an ordinary Contract with those for a Sectional Completion Contract in the one document, and it should be used in the same manner as the normal Building Contract and Scottish Supplement. Users of the new Scottish Building Contract for Sectional Completion are recommended to study these notes for guidance together with the notes on the 1980 Edition in the SBCC's Explanatory Memorandum.

It should be emphasised that this form is intended for use only when tenders have been notified that the Employer requires the Works to be carried out by phased sections each of which the Employer will take possession on practical completion of each section. It cannot be used for a Contract where the Works are not divided into sections in the tender documents; if by agreement with a Contractor an Employer wishes to take possession of a part of the Works and this has not been agreed in advance, then the provisions of Clause 18 of the Standard Form should be utilised.

The Sectional Completion Supplement divides the works into definite sections in the Contract documents with a corresponding division of the Contract Sum into section values and the fixing of separate completion periods for each section. The Contract remains, however, a single Contract and only one Final Certificate is issued at the end of the Contract; no provision is made for separate Final Certificates for each section. It is essential that the tender documents clearly identify the sections which together comprise the whole Works and these sections should be serially numbered and the appropriate numbers inserted in Appendix No. II of the Supplement. In this connection care should be taken in dealing with any part of the Works which is common to all or several sections (such as a boiler house serving 3 separate sections, each comprising a block of flats) to put this part of the Works into a separate section and to ascribe to it a Section Value and Date for possession, Date for Completion, Date of Liquidate Damages for delay and Defects Liability Period.

The following is a summary of the main modifications contained in the Sectional Completion Supplement:

1.   Clause 1.4 contains a definition of 'section' under which each section is identified by reference to the Contract Drawings and Contract Bills and to Appendix No. II.

2.   The Section Value ascribed to each section is entered in Appendix No. II for the purposes of Clauses 18 and 22A. 'Section value' is defined in Clause 18.1.6 as 'the value ascribed to the relevant section in Appendix No. II hereto' and the relevant

Appendix entry indicates that each section value is to be the total value of the section ascertained from the Contract Bills. The section values in toto must equal the Contract Sum and should take into account the apportionment of preliminaries and other like items priced in the Contract Bills.

3. There should also be entered in Appendix No. II, separately for each section, the Date of Possession, the Date for Completion, the Rate of Liquidate Damages for delay and the Defects Liability Periods.

4. The Contractor is to be given possession of each section on the date shown in Appendix No. II and should proceed with each section concurrently or successively as required by the Contract and must carry out each Section within the Contract period stated in the Appendix or any extended period authorised by the Architect. Liquidate damages for delay for each section are separately calculated where there is delay in completing any section.

5. On practical completion of any section the Architect must issue a Practical Completion Certificate for that section (Clause 17.1). Consequently, the Contractor is relieved of his corresponding duty to insure that section under Clause 22A if applicable and the first half of the retention money attributable to that section must be released (Clause 30.4). A separate Defects Liability Period also operates for each section (Clause 17.2).

6. When all the sections have been completed the Architect issues a Certificate to that effect (Clause 17.6) and the period begins to run within which the final Account is to be prepared and the Final Certificate issued for the whole of the Works comprising all the sections (Clause 30.6).

7. The Architect must note that when making insurance arrangements under Clause 21.2 agreement must be reached with the Insurers as to whether any section for which a Practical Completion Certificate has been issued is to be treated as continuing to be included in the Works and so not insured, or is to be treated as 'property other than the Works' and so covered by the insurance.

Explanatory notes

Clause 1   Here either Bills of Quantities or Specification should have been deleted.

Clause 2   Here the alternative forms of the Conditions are deleted to leave the Standard Form which is operative for this Contract.

Clause 3   There are four references where either Architect or Supervizing Officer should be deleted.

Clause 6   This is the Attestation where the words at the beginning of the clause are either left in or deleted. If left in then the Contract is liable for stamp duty. Again, it is suggested that legal advice should be sought on this matter.

Scottish Supplement for Sectional Completion

Clause 5   Here a deletion should be made to establish whether or not Clause 5.3.1.2 is operative. This refers to the Contractor's Master Programme.

Clause 17.6   Either Architect or Supervizing Officer should be struck out.

Clause 22   Only one of the three alternatives 22A, 22B or 22C should be left in and it should be noted at whose risk the insurance has to be effected.

Clause 25.3.6   Again a reference to Architect/Supervizing Officer.

Clause 30   Reference to Architect/Supervizing Officer.

Appendix II

Clause 31   Here the status of the Employer is ascertained.

Clause 37   Two alternatives to the fluctuation clause should be deleted. The percentage addition should refer to the applicable clause. Signatures are required on Pages 3, 7 and 8.

Chapter 2

General

Clause 1 – Interpretation, definitions, etc.
England and Wales

Here is given the instruction to read the Articles of Agreement, the Conditions and the Appendix as one document, or as a whole. Unless there is a specific provision to the contrary, reference to 'clauses' means the clauses of the conditions. Clause 1 goes further in stating, again subject to any provision to the contrary, that the effect or operation of any article, clause, item or entry will be governed by any relevant qualification or modification in any other article, clause, item or entry in any of the parts making up the whole. The parts being the Articles of Agreement, the Conditions and the Appendix.

The remainder of this clause is an alphabetical directory of definitions of words or phrases contained in the parts. The relevant article number, appendix entry or recital number is appended for reference. Where the definition refers to one clause only, then the clause number is stated where that definition is set out.

It should be noted that this is the only clause which sets out and defines the Perils in Clause 22 (one of the three insurance clauses) and the exclusions thereto.

Scotland

It should be noted that in Scotland there are two appendices to the Scottish Building Contract. The first is the Scottish Supplement and the second, Appendix II, is the equivalent to the Appendix to the Conditions of the Standard Form in England and Wales.

The modifications to Clause 1 to bring it into line in Scotland are in the form of deletions and additions as set out in the Scottish Supplement which forms part of the Scottish Building Contract between the Employer and the Contractor. These are set out in full below.

Scottish Supplement

Deletions	Additions
Clause 1.1 and 1.2 wholly deleted	
Clause 1.3 part deleted as under	Clause 1.3 part added as under
Appendix	Appendix II to the Building Contract
Arbitrator	Arbiter
Articles or Articles of Agreement	The foregoing Building Contract
Contract Bills	The Bills of Quantities referred to in the Building Contract which have been priced by the Contractor and signed by him and by the Employer or on their behalf
Contract Drawings	The drawings referred to in the Building Contract which have been signed by the Employer and the Contractor or on their behalf
Contract Sum	The sum stated in the Building Contract or such other sum as becomes payable in accordance with the Conditions subject to Clause 15.2

Nominated Sub-contract Documents

Deletions	Additions
Tender NSC/1	Tender NSC/1/Scot
Agreement NSC/2	Agreement NSC/2/Scot
Agreement NSC/2a	Agreement NSC/2a/Scot
Nomination NSC/3	Nomination NSC/3/Scot
Sub-contract NSC/4	Sub-contract NSC/4/Scot
Sub-contract NSC/4a	Sub-contract NSC/4a/Scot
Works	The works described in the Building Contract and shown and described on Contract Drawing and in Contract Bills
The following Clause 1.4 to be added	
1.4 Additional definitions(*see NB*)	
Execution of this Contract (5.2 and 5.3)	Formal adoption and signing of the Building Contract
Execution of a binding Sub-contract Agreement (19.3)	Creation of a Sub-contract
Article 3A, 3B and 4	Clause 3 of the Building Contract
Article 5	Clause 4 of the Building contract
Real or personal	Heritable or moveable
Section 117, Local Government Act 1972	Section 68, Local Government (Scotland) Act 1973

NB: Clause 1.4 changes certain words and phrases and their meanings to enable the Conditions to be used in Scotland. Accordingly, the changes to these words, phrases and meanings, wherever they occur throughout all the Clauses of the Conditions, will be deemed to have taken place and will not be referred to again.

Chapter 3

Responsibilities after execution of the Contract

Clause 2 – Contractor's obligations
Clause 4 – Architect's instructions
England and Wales

Clause 2 – Contractor's obligations

2.1   This clause reinforces the point that this is an entire contract and the Contractor's basic obligation is to carry out and *complete* the works as described in the Contract Documents.

The Contract Documents are defined here as

(a) The Contract Drawings,
(b) The Contract Bills,
(c) The Articles of Agreement,   Collectively
(d) The Conditions,   called the 'Contract
(e) The Appendix.   Documents'

The Contractor has only to produce what is set out in the Contract Documents and therefore has no responsibility for any design of the works.

The quality of materials and standard of workmanship must meet the specification of the Contract Documents. These standards will most likely be contained in the Contract Bills, but other specification data might be shown on the Drawings.

The Architect is allowed here to exercise his opinion on the quality of materials and standard of workmanship, and the quality and standard are to be to his reasonable satisfaction.

However, in expressing his opinion, the Architect cannot impose upon the Contractor obligations greater than those set out in the Contract Documents. Reasonable satisfaction could be subject to arbitration if the Contractor feels that the Architect has been unduly biased on any point in his decision.

2.2.1   It is clearly stated here that the Contract Bills should not increase, vary or change in any way the obligations of the parties as they are set out in the other documents, namely the Articles of Agreement, the

Conditions and the Appendix. If this situation arose, then the Contract Bills would not take precedence nor have any effect over the other documents.

2.2.2   The Contract Bills have to be prepared in accordance with SMM6. Where they have either not been in their entirety or for groups of items, or for any item, then the basis of that preparation must be stated specifically.
  If the Contract Bills

(a) Are not in accordance with SMM6, or
(b) Are not in accordance with the basis of preparation as specifically stated, or
(c) Have errors in description, or
(d) Have errors in quantities, or
(e) Have items omitted,

then the Contract is not invalidated or made ineffective, but the Contract Bills have to be corrected as if they were a variation under Clause 13.2. This correction will take place regardless of whether it is an increase to, or a decrease from the Contract Sum. This sub-section does not specifically lay the onus upon the Contractor to point out to the Architect any of these anomalies, but it would be in his own interest to do so.

2.3   Here it is clearly the Contractor's obligation to give the Architect notice in writing of any discrepancies between two or more documents or parts of documents, but only if he 'shall find' such discrepancies; he is not expected to carry out an exhaustive examination for this purpose.
  The listed documents are

(a) The Contract Drawings,
(b) The Contract Bills,
(c) Any instruction issued by the Architect under the Conditions (except an instruction which requires a variation under Clause 13.2), and
(d) Any drawings or documents issued by the Architect under Clauses 5.4 or 5.7.

The Contractor's notice is required *in writing* and it is of paramount importance that he carries out the procedures of the Conditions exactly. This could affect any future claims he might make for extension of time or possible claims for loss and expense.
  The Architect now has the duty to issue instructions with regard to the discrepancy, clearly indicating which documents will stand and which will be corrected.
  The further inference could be made that if the Architect or Quantity Surveyor discovers such a discrepancy, then the procedure for correction would follow without the Contractor's initial instigation. In that case, the Contractor's written confirmation of the instruction would help to protect his future interests.

Clause 4 – Architect's instructions

4.1.1   The Contractor must comply with all instructions from the Architect provided that the latter is empowered by the Conditions to issue the instruction. (*See* at the end of this clause the list of clauses which empower the Architect.)

There is one exception when the instruction is a variation under Clause 13.1.2. This clause refers to any addition, alteration or omission of any obligations or restrictions imposed by the Employer in the Contract Bills, concerning

(a) Access to the site or use of any specific parts of the site,
(b) Limitations of working space,
(c) Limitations of working hours, and
(d) The execution or completion of work in any specific order.

All these matters will in all probability have been detailed in the Preliminaries Bill.

The Contractor can object to an instruction which covers matters (a) to (d) above, but he must be reasonable in his objection and must state the objection in writing to the Architect.

4.1.2   If, after a written notice from the Architect requiring the Contractor to comply with an instruction, he does not comply within seven days (seven days, not a week), then the Employer can employ and pay other persons to do the work. All costs connected can be either deducted from monies due to the Contractor or recovered from him as a debt.

At first sight it would appear that the Employer is getting the work done for nothing, but obviously all costs are those over and above the payment that the Contractor would have received if he had carried out the work in the instruction. Care is needed here since cost might well include additional professional fees incurred by the Architect and the Quantity Surveyor. In most cases, ignoring an Architect's written instruction, or deliberately neglecting to carry it out, will hardly engender the rapport and harmonious working relationship that it is helpful to create.

4.2   Should the Contractor have any doubts as to the Architect's power to issue an instruction under the Conditions, he can request the Architect to specify the Clause on which he has based his power to instruct. Such request must be in writing and the Architect must reply in writing. The Conditions state that the Architect shall forthwith comply with the request, but the reasonable Contractor would not expect a reply literally 'by return of post'. The Architect's reply will state the clause under which the matter is being instructed and this will clarify for the Contractor whether it will be a variation which varies the Contract Sum, or work to be done at his own expense.

After receiving the Architect's reply, the Contractor will either obey the instruction and thus signify acceptance of the reply, or dispute it and go to arbitration. This is a case where arbitration can be instigated immediately and need not wait until the end of the Contract.

In the Conditions the words 'deemed for all the purposes of this Contract' mean that by complying with the Architect's instruction, the Contractor has effectively barred not only his own chance for arbitration over the correctness of the instruction but also the Employer's ability to go to arbitration on the same point.

4.3   The Architect should give all his instructions in writing. However, since oral instructions are often necessary and cannot be ignored, the Conditions set out a standard procedure to be followed.

On receipt of an oral instruction the Contractor has seven days within which to confirm the instruction in writing. Within the following seven-day period, if the Architect has not dissented from this written confirmation, then the confirmation stands as if it had been an Architect's written instruction.

Two other alternatives are also catered for:

(a) If the Architect himself confirms his oral instruction within seven days then the instruction is dated as the confirmation and the Contractor need do nothing further.

(b) If neither the Architect nor the Contractor confirms an oral instruction and the latter carries out the work of the instruction, the Architect can confirm in writing at any time up to the issue of the Final Certificate. It would be wise for the Contractor to obtain confirmation as soon as possible while the matter is still fresh in everyone's mind. Note that in this case, the date of the instruction will be when it was issued orally, and not the date of the written confirmation.

Architect's Instructions are empowered, in accordance with Clause 4, under the following clauses:

2.3   In regard to discrepancies in or between documents.

6.3   In relation to divergence between statutory requirements and documents and variations.

7   In regard to errors in setting out.

8.3   Requiring opening up for inspection and testing.

8.4   In regard to removal of work and materials not in accordance with the Contract.

8.5   Requiring exclusion from the works of any person employed.

13.2   Requiring a variation.

13.3.1   In regard to expenditure of Provisional Sums in Contract Bills.

13.3.2   In regard to expenditure of Provisional Sums in a sub-contract.

17.2   In a Schedule of Defects as an instruction.

17.3   (When considered necessary) requiring defects, etc. to be made good.

22C.2.3.2   Requiring the Contractor to remove debris after fire, or other, damage.

23.2	In regard to the postponement of any work.
32.2	Requiring protective work on outbreak of hostilities.
33.1.2	Requiring removal or disposal of debris or damaged work and the execution of protective work in connection with war damage.
34.2	In regard to what is to be done with alleged fossils, antiquities, etc.
35.10.2	Nominating proposed sub-contractor (Basic Method).
35.11.2	Nominating proposed sub-contractor (Alternative Method).
35.23	Requiring the omission of work for which he intended to nominate a proposed sub-contractor, when the nomination has failed.
35.24	Specifying default by a nominated sub-contractor.
36.2	For the purpose of nominating a supplier.

Scotland

Clause 2 – Contractor's obligations

There is no change to this clause for use in Scotland. It is worth noting, however, that because of Clause 1, the Contract Documents will be defined as:

(a) The Contract Drawings,
(b) The Contract Bills,
(c) The Scottish Building Contract (not Articles of Agreement),
(d) The Conditions (amended by the Scottish Supplement), and
(e) Appendix II.

Clause 4 – Architect's Instructions

Again, there is no change in this clause for use in Scotland.

Chapter 4

The Contract Sum and Contract Documents

Clause 3 – Contract Sum – additions or deductions – adjustment – Interim Certificates

Clause 5 – Contract Documents – other documents – issue of certificates

Clause 14 – Contract Sum

England and Wales

Clause 3 – Contract Sum – additions or deductions – adjustment – Interim Certificates

In various clauses throughout the Conditions there are facilities for adjusting, adding to, or deducting from the Contract Sum. This clause provides that when any such amendments are valued, in whole or in part, they shall be taken into account in the next Interim Certificate.

This then ensures that all proper adjustment of monies takes place as the Contract proceeds and the Contractor's 'cash-flow' situation is more easily maintained. However, the words 'as soon as such amount is ascertained in whole or in part', presuppose some liability upon the Contractor to be helpful to all parties concerned, to obtain an agreed amount in a reasonable time. If he is unhelpful and slow to agree to an amount that is to be omitted from the Contract Sum, he can hardly expect a speedy conclusion in the opposite case.

Clause 5 – Contract Documents – other documents – issue of certificates

5.1   The Contract Drawings and the Contract Bills, after being signed in relation to the Contract, are kept in the custody of either the Architect or the Quantity Surveyor. They are available at all reasonable times (business hours) for inspection by the Employer and the Contractor.

5.2

5.3   Further obligations are placed upon both parties to provide documents which are listed below with their respective responsibilities:

5.4

Contractor to Architect	Architect to Contractor
Two copies of the Master Programme and within 14 days amended copies relative to decisions under Clauses 25.3.1 or 33.1.3	One copy of the signed Contract Documents (these are listed in Clause 2 and should be signed and certified as true copies)
	Two further copies of the Contract Drawings
	Two copies of the unpriced Bill of Quantities
	Two copies of any descriptive schedules or like documents
	Two copies of any further drawings or details necessary

With regard to documents it should be noted that the descriptive schedules, or like documents, and the Contractor's Master Programme, will not impose any further obligations beyond those in the Contract Documents. The Contractor's Master Programme has no laid-down format, unless one was to be stated in the Bill of Quantities or Drawings, and it would appear that the Contractor has no further responsibility to produce anything beyond that which he would normally prepare for his own use, e.g. a bar chart. Failure by the Contractor to keep his Master Programme up to date or failure to provide it in the first place, would be a breach of contract, but only if it had been required in the Bills.

5.5 The Contractor is required to keep on site one copy of the following documents for the use of the Architect or his representative:

(a) One copy of the unpriced Bill of Quantities,
(b) One copy of the descriptive schedules and other like documents,
(c) One copy of the Master Programme,
(d) One copy of any further drawings or details, and
(e) One copy of the Contract Drawings.

5.6 There is a clear duty on the Contractor to return all drawings and documents which bear the Architect's name, after the balance of the Final Certificate has been paid, but only if the Contractor is so requested.

5.7 The Contractor must not use any of the documents in this Contract for any purpose other than the Contract, this being, if you like, a protection of copyright to the Architect. Equally, neither the Architect nor the Quantity Surveyor will divulge any rates or prices in the Contractor's Bills of Quantities, except for the purpose of this Contract.

5.8 This last sub-clause provides that all certificates issued by the Architect will be to the Employer, with a copy to the Contractor. There is one exception to this rule, and that is the 'Certificate of failure to complete Nominated Sub-contract Works' (Clause 35.15.1), which is issued directly to the Contractor.

The following list is of certificates which the Architect is required to issue from time to time, with the relevant clause numbers appended.

Architect's Certificates required under the provisions of the Standard Form of Building Contract

17.1   Certificate of Practical Completion

17.4   Certificate of Completion of Making Good Defects

18.1   Certificate of Value of Relevant Part (Partial Possession)

24.1   Certificate of Non-completion

30.1   Interim Certificate

30.8   Final Certificate

35.13.5.2   Certificate that the Contractor has failed to provide reasonable proof of discharge of amounts directed to the Contractor of interim or final payment to the Nominated Sub-contractor

35.15.1*   Certificate of Failure to Complete Nominated Sub-contract Works

35.16   Certificate of Practical Completion of Nominated Sub-contract works

> *This is the only certificate which is the exception to the general rule of Clause 5 that certificates be issued to the Employer. The Clause 35.15.1 certificate is to be issued directly to the Contractor.

Clause 14 – Contract Sum

14.1   The quality and quantity of work in the Contract Sum is only that contained within the Contract Bills, not the Drawings; the Bills, therefore, are the basis from which instructions and variations vary quality or quantity of the work. (*See also* 2.1).

14.2   The Contract Sum will not be adjusted by either party to correct any errors, or whatever, in the Contractor's pricing, calculating, totalling or collection of monies. The Contract Sum in this regard is inviolate. Clause 2.2.2.2 could, however, raise a variation which would correct the Contract Sum, if there were to be a departure from the method of preparation referred to, an error in description, or in quantity or omission of items. (*See* Chapter 3.)

The only way that the Contract Sum can be adjusted or altered, apart from Clause 2.2.2.2 above, is by express provisions in the Conditions, which are as follows:

6.2   Addition of fees and charges,

7   Architect's Instruction to adjust for amending errors in contractor's setting out,

8.3   Addition of the cost of opening up, testing and making good when materials and goods are found to be satisfactory,

9.2   Additions of royalties, etc. arising out of Instructions,

13.7   Addition or deduction from valuation of variations (*See* Clauses 2.2.2.2 and 6.1.4.3 for additional variations),

17.2 ⎱ Architect's Instructions for adjustment from making good
17.3 ⎰ defects, and

21.2.3 Additions from maintaining insurances, Clause 21.2.

It should be noted that the Contract Sum is the amount in the Articles of Agreement signed by both parties. It is possible for the tender document to have an agreement added whereby the Contractor will agree to either Alternative 1 or Alternative 2 being applied, as set out in the Code of Procedure for Single Stage Selective Tendering 1977. This can have the effect of adjusting the Tender Amount for arithmetical or pricing errors, or endorsing, in which case the Contract Sum might not be the same as the amount in the tender. The wording in Sub-clause 14.2 applies after the Contract Sum has been established.

Scotland

Clause 3 – Contract Sum – additions or deductions – adjustment – Interim Certificates

There are no changes to this clause for use in Scotland.

Clause 5 – Contract Documents – other documents – issue of certificates

The Scottish Supplement provides (on Page 4) two options, one of which must be deleted. The options refer to the Contractor's Master Programme.

The first option leaves in Clause 5.3.1.2 which requires the Contractor to provide a Master Programme, and duly amend it as required.

The second option deletes the clause and absolves the Contractor of the liability to produce a Master Programme. It refers to Clause 5.3.2 from which shall be deleted the words 'not in the master programme for the execution of the works or any amendment to that programme or revision therein referred to in Clause 5.3.1.2'.

This cross-checks back to the Scottish Supplement of the Building Contract and is one of the alternatives you must note before signing a contract.

Clause 14 – Contract Sum

The Scottish Supplement amends this clause in 14.2, second line, by adding after the word 'Conditions' the words 'including without prejudice thereto Clause 30.11'.

Clause 30.11 refers to the addition or deduction of Liquidated and Ascertained Damages from any sum due to the Contractor or by the Contractor to the Employer, as the case may be, under the Final Certificate. This is an extra clause added by the Scottish Supplement.

Chapter 5

Possession of the site

Clause 6 – **Statutory obligations, notices, fees and charges**

Clause 7 – **Levels and setting out the works**

Clause 8 – **Materials, goods and workmanship to conform to description, testing and inspection**

Clause 9 – **Royalties and patent rights**

Clause 23 – **Date of possession, completion and postponement**

England and Wales

Clause 6 – Statutory obligations, notices, fees and charges

6.1 The obligation is imposed upon the Contractor to comply with all the relevant legislation, bye-laws, building regulations and other statutory requirements and to give all the necessary notices. This will include any local authority or statutory undertaker who is involved in the works. Examples of these are connections to sewers, gas mains, water mains and electricity supplies, etc.

If the Contractor finds any difference between the statutory requirements and either all of the documents or any one document, or any instruction requiring a variation, then he must give the Architect a written notice pointing out the difference or differences. If, in his turn, the Architect discovers a difference himself, or receives a notice from the Contractor pointing out a difference, he will in either case issue an instruction within seven days in relation to the difference. If the instruction requires the works to be varied, it is then treated as an instruction requiring a variation.

Should an emergency arise the Contractor can carry out the minimum work necessary, before receiving an instruction, in order to comply immediately with a statutory requirement. He will inform the Architect of the emergency and the steps he is taking to deal with the situation. The Architect should make sure than an emergency really exists before proceeding. Again, this work will be treated as a variation because of an Architect's Instruction.

In Sub-clause 6.1.5, provided that the Contractor has given written notice of a difference, he is then not liable if the works do not accord with statutory requirements, because he has carried out the work in accordance with the documents or a variation.

6.2   The Contractor will pay and indemnify the Employer for any liability in respect of all fees and charges, including rates or taxes, which are legally demandable by a statutory undertaker in respect of the works, less any VAT. In turn, the Contractor will recover the monies paid out by adding them to the Contract Sum, provided that the recovery of the monies is not allowed for elsewhere. The Contractor would be reimbursed elsewhere if:

(a) The work or materials or goods is done or supplied by a local authority or statutory undertaker acting as a Nominated Sub-contractor or a Nominated Supplier,

(b) They are priced in the Contract Bills, or

(c) They are a Provisional Sum in the Contract Bills.

In essence, therefore, the Contractor will be reimbursed in one form or another for all legally-demanded fees and charges which he has been required to pay.

6.3   Clause 19 (Assignments and Sub-contracts) and Clause 35 (Nominated Sub-contractors) do not apply to work by a local authority or statutory undertaker who execute work only because of statutory obligations. In the terms of the contract they are neither domestic nor nominated sub-contractors.

As a cross-reference to Clause 25 (Extension of time) a delay could be claimed as set out in Clause 25.4.11 with regard to local authorities or statutory undertakers, but not direct loss and expense in Clause 26.

Clause 7 – Levels and setting out the works

The Architect has the sole responsibility of furnishing the Contractor with the levels and dimensional drawings to enable him to set out the works properly and to the required levels. It is the Contractor's sole responsibility, using this information, to set out and level the works accurately and also be responsible for amending his own errors at his own cost.

However, the Architect is empowered to instruct that all or part of the cost of erroneous setting out can be added to the Contract Sum. On the face of it, it would appear that the Employer is being asked to pay for the Contractor's mistakes. Although the meaning is not clear, it would appear likely that the Architect would issue an instruction with regard to re-setting-out only if he was in error, in part or in whole, with his information to the Contractor.

Clause 8 – Materials, goods and workmanship to conform to description, testing and inspection

8.1   The Contractor's standards for workmanship, materials and goods are set out in the Contract Bills, and as far as can be obtained they will be adhered to as his responsibility. The descriptions in the Bill items and the descriptive preambles or specifications should be read in conjunction to ensure complete understanding of what is required. Notice that any conflict between the two sources of information would need a written notice from the Contractor, as described in Clause 2.3.

If for any reason, the Contractor is unable to supply materials or goods described in the Contract Bills, again an immediate written report to the Architect is necessary. The Contractor can suggest alternatives but not make a substitution without permission.

It is understood that work not in the Contract Bills, but which might appear in a variation will also have the same standards applied.

8.2   The Architect has the power to request proof of the standard of the materials and goods by the production, by the Contractor, of his vouchers, delivery notes or invoices.

8.3   The Architect has the further power to instruct the Contractor to open up work for inspection, or to perform tests and afterwards make good. If his standard of workmanship and materials and goods, meets that which is required by the Bills, the whole cost will be added to the Contract Sum and the Contractor therefore recovers the calculated amount of compliance. On the other hand, should the opening up or tests show that the Contractor is not consistant with the Standards required, he will bear the entire cost together with the additional cost of upgrading his workmanship, materials or goods to the proper standards.

It is worth remembering that if the Contractor is proved to be in the right, in the above, then this is one of the grounds for extension of time under Clause 25, and a ground for a claim for loss and expense under Clause 26.

8.4   The Contractor must comply with instructions for the removal of work or materials not in accordance with the contract, which does not affect his contractual liability to carry out and complete the works.

Failure to carry out such an instruction, is a ground for determination by the Employer under Clause 27.1.3.

8.5   The Architect has the power to instruct the exclusion from the works of any person employed on the works. He is required not to be unreasonable or vexatious in his decision. 'Person' in this context is probably intended to be singular but regard the definition in Clause 1.3.

Clause 9 – Royalties and patent rights

9.1   The Contractor must indemnify the Employer against any kind of claim arising out of his infringement of patent rights on materials, goods, processes, etc., described in the Contract Bills, and he is deemed to have included all his outlays in respect of royalties, etc., in his tender.

9.2   If, because of an Architect's Instruction the Contractor is involved in the payment of royalties, etc., he will be reimbursed by an adjustment to the Contract Sum and relieved of his indemnification liability against infringement of patent rights to materials, goods, processess, etc., in that instruction.

Note that this clause could involve the Contractor in damages to be paid for infringement as in Clause 9.1, but these damage payments would also be added to the Contract Sum, as in the case of Clause 9.2.

Clause 23 – Date of possession, completion and postponement

23.1   It is worth restating that the dates for both possession and completion will be entered in the Appendix. It is of course important that the Contractor takes note of both dates and therefore the time between, this being his contract period.

The Contractor will be given possession of the site on the date entered in the Appendix, the site boundaries being clearly marked on a contract drawing or described in the Preliminaries Bill. He is then expressly obligated to proceed regularly and diligently with the works and complete on or before the completion date, or such other completion date as may be fixed under Clause 25 or 33.1.3. His Master Programme will give the Architect a reasonable guide as to his performance and whether or not his progress is likely to meet the completion date.

Failure on the Contractors part to proceed could lead to determination by the Employer under Clause 27, but this would need either a complete breakdown of activity or such slow progress as would make it impossible for the Contractor to complete in a reasonable time.

On the other hand, if the Contractor is not given possession of the site on the stated date, the Employer is in breach of contract. Would it be to his advantage to have his contract of employment terminated at this point? Perhaps one course of action would be to request an instruction from the Architect for postponement under Clause 23.2, and follow this up with a claim for extension of time under Clause 25 with a further claim for loss and expense under Clause 26. Of course, if such a postponment carries on continuously in excess of the period of delay stated in the Appendix, the Contractor has no option but to determine the contract. Another option is to negotiate partial possession of the site.

The stated Completion Date forms the starting point for extension of time under Clause 25 and loss and expense under Clause 26. Remember that completion would mean having the works in such a state as would allow the Architect to issue his Certificate of Practical Completion.

23.2   The Architect in this sub-clause is given the power to postpone the whole or part of the works and, as stated before, it enables the Contractor to claim for an extension of time and loss and expense, but not to go to arbitration. Further, if the postponement exceeds the period of delay stated in the Appendix, the Contractor can raise determination proceedings.

Scotland

Clause 6 – Statutory obligations, notices, fees and charges

Clause 7 – Levels and setting out the works

Clause 8 – Materials, goods and workmanship to conform to description, testing and inspection

Clause 9 – Royalties and patent rights

There are no changes to the above clauses for use in Scotland.

Clause 23 – Date of possession, completion and postponment

Although there is no change in this clause for use in Scotland, it is worth restating that the relevant dates and periods of delay are to be found in Appendix II in the Scottish Building Contract.

Chapter 6

Site supervision

Clause 10 – Person-in-charge

Clause 11 – Access for the Architect to the Works

Clause 12 – Clerk of Works

England and Wales

Clause 10 – Person-in-charge

The Contractor is required to have someone on site at all times, probably 'during normal working hours' would define this. The person will of course vary from contract to contract dependent on the size and the complexity of the Contractor's organization. A foreman on one contract might be sufficient, whereas only a site agent or contract manager would suffice on another. The two criteria to satisfy are (i) the person appointed be constantly on site, and (ii) that person be capable of receiving, and authorized by the Contractor to receive, Architect's Instructions and Clerk of Works' directions and be able to carry them out on the Contractor's behalf.

Note that the Architect could issue an instruction to the Contractor's head office, the Person-in-charge not being the exclusive recipient of instructions because he is on the site. However, when on site the Architect will not give anyone else instructions unless there is a specific delegation from the Contractor's Person-in-charge.

In his own interests, it would be prudent of the Contractor to let the Architect know who he is placing in charge on the site, perhaps introducing him either at the first meeting or in writing to the Architect. Remember that any change of the Person-in-charge should also be communicated to the Architect.

Clause 11 – Access for the Architect to the Works

The Contractor has to grant to the Architect and his representatives right of access to the site, his own workshops and any other place where work is being prepared for the contract. He has further to secure right of access for them to his domestic sub-contractors' workshops or any other place where they are preparing work for the contract, and also the same in respect of nominated sub-contractors, but note the words 'as far as possible'. The Architect will be reasonable in the exercising of his rights in this respect.

The right of access is 'at all reasonable times' which would be construed to mean business hours.

It will be found in other chapters of this book that both nominated sub-contracts and domestic sub-contracts contain clauses giving the right of access as explained above. The formation of these contracts, therefore, will place the same obligation upon them.

Clause 12 – Clerk of Works

The Employer can appoint either a Clerk of Works or the Architect as his professional advisor, but a clerk does not become the Architect's representative on site, merely an inspector. The Clerk of Works can only give directions about the same matters as the Architect is empowered to issue instructions.

The Contractor must offer the Clerk of Works reasonable facility to carry out his duties. In practice, this would mean free access to inspect the works and a degree of co-operation from the Contractor would obviously promote greater harmony on site.

It must be clearly understood that the Clerk of Works cannot issue an instruction, only a direction. This direction is not valid and can be ignored unless the Architect confirms it in writing within two working days, when it then becomes an Architect's Instruction dated as the date of the confirmation.

The Contractor can confirm a Clerk of Works' direction to the Architect, but this still does not make it valid. However, it is good practice on the Contractor's part to do so and thus request the Architect's confirmation of the direction within two working days.

What happens when the Architect goes beyond two working days before issuing his confirmation does not appear to be covered by this clause. Only if the direction subsequently becomes a variation under Clause 13 will the matter be resolved, but not otherwise. It would appear that confirmation in excess of two working days would render the direction invalid, but the Contractor's discretion with regard to the time scale would be best exercised.

Scotland

Clause 10 – Person-in-charge

Clause 11 – Access for the Architect to the Works

Clause 12 – Clerk of Works

There are no changes to the above clauses for use in Scotland.

Chapter 7

Variations to the works

England and Wales

Clause 13 – Variations and Provisional Sums

As has been seen previously, the Architect can issue instructions, which in turn have to be empowered under the Conditions, but not all instructions become variations which will adjust the Contract Sum. Clause 13.1 defines under two sub-clause numbers what constitutes a variation, as follows:

13.1.1   The alteration or modification of the design, quality or quantity of the works as shown upon the Contract Drawings and described by or referred to in the Contract Bills; including

the addition, omission or substitution of any work,

the alteration of the kind or standard of any of the materials or goods to be used in the works, and

the removal from site of any work executed or materials or goods brought thereon by the Contractor for the purposes of the works (other than work, materials or goods which are not up to standard).

Note that in Chapter 5 Clause 8 the kinds and standards are stated to have been described in the Contract Bills. Also, the removal of sub-standard work and materials and goods is covered, and their removal does not become a variation, as stated in the parentheses above.

13.1.2   The addition, alteration or omission of any obligations or restrictions imposed by the Employer in the Contract Bills in regard to

access to the site or use of any specific parts of the site,

limitations of working space,

limitations of working hours, and

the execution or completion of the work in any specific order.

The Contractor will notice that this Sub-clause deals only with variations to restrictions or obligations imposed by the Employer and could, of course, be additional restrictions or obligations. As a cross-check to Clause 4, the Contractor can make reasonable objections in writing to compliance with such a variation. He must also look to Clause 25 (extension of time) and Clause 26 (loss and expense) if he was denied access to the site, which is one of the four headings empowering this type of variation.

13.1.3   The substitution of measured quantities duly priced by the Contractor in the Contract Bills, by a nominated sub-contract, is NOT a variation. If the Architect wished to make this type of change, he would require the Contractor's agreement, and presumably he would state his own terms.

13.2   In addition to the Architect issuing instructions which require a variation, under this sub-clause he can sanction a variation carried out by the Contractor without Instruction. The latter would, however, be very imprudent to carry out any variation on his own authority without first requesting an instruction, in writing, from the Architect. His sanction would, of course, have to be after the work was carried out, in other words, retrospective. However, this must not be confused with the provisions in Clause 6 where emergency work may have to be carried out to meet statutory requirements.

Other variations may arise because of provisions in Clauses 2.2.2.2, 6.1.3 and 13.5.5 where matters can be treated as if they were variations required by an instruction. These are other sources of variations and they become variations which can vary the Contract Sum, the same as any other.

13.3   Lastly, the Architect will issue instructions as to how Provisional Sums in the Contract Bills and Sub-contracts are to be expended. It is possible that this could be a part or complete omission, or possibly a Prime Cost Sum could arise out of the original Provisional Sum. Definitions of Provisional and Prime Cost Sums are to be found in SMM6.

No variation can vitiate the Contract and it is really the Architect's province to ensure that any variation is not so extensive as to become such a major change that will endanger this provision. This is a point that the Contractor should also be taking into consideration.

13.4   This sub-clause really deals with exceptions to how the variations are valued, and sets out definite rules of procedure. The valuation is to be carried out exclusively by the Quantity Surveyor named in the Contract Documents.

The rules for valuation can be waived, if there is any other agreement between the Employer and the Contractor as to how the variations will be valued. This is an unlikely circumstance, as the Conditions would require amendment to accomplish the change. Note that the Quantity Surveyor has no authority to change the rules.

The rules for valuing variations to works done by a nominated sub-contractor are contained in Sub-contract NSC/4 or NSC/4a as applicable, unless otherwise agreed by the Employer, Nominated

Sub-contractor and Contractor, and are dealt with in a subsequent chapter.

An Architect's Instruction on the expenditure of a Provisional Sum can raise a Prime Cost Sum. Under Clause 35.2 the Contractor can tender for the work in the Prime Cost Sum. If he is successful and accepted, then that work is valued at the acceptance figure and not included in the valuation of the original instruction. This is explained in greater detail in a subsequent chapter, but merely stands here as an exception to the normal rules of valuation of a variation.

13.5   The procedure for valuing variations and instruction as to the expenditure of Provisional Sums is summarized as follows:

Type of work	How valued
(i) work which can be properly measured (ii) of a similar character (iii) done under similar conditions (iv) not a significant change in quantity	Rates and prices in the Contract Bills
(i) work which can be properly measured (ii) of a similar character (iii) not done under similar conditions (iv) and/or a significant change in quantity	Rates and prices in the Contract Bills with a fair allowance for difference in conditions and/or quantity
(i) work which can be properly measured (ii) not of a similar nature	Fair rates and prices
(i) work which cannot be properly measured	Daywork at the rates in the Contract Bills or for any specialist trade
(i) work omitted as set out in the Contract Bills	Rates and prices as set out in the Contract Bills (*see also* Clause 13.5.5)

It will be noticed that the basis for all valuation in this clause is the Contract Bills and the rates and prices contained therein, and that the rules are applied in succession until the Contractor has a set that fits the particular case. The Quantity Surveyor has no right to examine your rates as to their profit element or whether or not they are economic, but if you disagree with the rate being used in the valuation, you may find that a 'break-down' of a rate is necessary to obtain a fair *pro-rata* rate.

The Quantity Surveyor has to exercise his judgment on these four main points

(a) Can the work be properly measured?
(b) Is it of a similar character?
(c) Is it done under similar conditions?
(d) Is there a significant change in quantity?

His decisions will have to take into account a large number of factors and, in practical terms, the Contractor will probably have made up his own list of reasons which in his opinion could affect the decision. He can, of course, have discussion with the Quantity Surveyor and should they fail to agree, arbitration is their recourse.

The Quantity Surveyor shall, in his valuation, use the same principle of measurement as was used in the Contract Bills.

An allowance for any percentage or lump sum adjustments to the Contract Bills will also be made in the valuation.

Where necessary, adjustment of Preliminaries Bill items will be made.

When the Quantity Surveyor has decided that measurement is not appropriate to value a variation, the rules state that daywork will be allowed. Daywork will be allowed on an agreed definition of Prime Cost of Daywork with the normal percentage additions as filled in by the Contractor in the Contract Bills. This agreed definition of Prime Cost of Daywork will be the current issue, at date of tender, as issued by the RICS and the NFBTE.

When the work is in a specialist trade, the current, at date of Tender, agreed definition of Prime Cost of Daywork will be used for that trade with percentage additions. The RICS have agreed these definitions with the Electrical Contractors' Association and the Heating and Ventilating Contractors' Association.

The Conditions set out the procedure for the submission of vouchers which show daily time spent, workmen's names, and plant and materials used. The vouchers to be submitted to the Architect or his representative not later than the end of the week following that in which the work was done. Although it is technically wrong not to carry out the procedure to the letter, it is difficult to envisage a Quantity Surveyor, who knows that the valuation of a variation can be valued only as daywork, expecting to receive the vouchers within a week of the work being done. A fair and reasonable approach should be taken by all parties.

Because of a variation or an instruction for the expenditure of a Provisional Sum, other directly-associated work may have its Conditions changed substantially. Accordingly, that other work will also be treated as if it were a variation and so valued in accordance with the foregoing rules. This could present a problem in definition, and the Contractor will be well advised to examine each variation or instruction for the expenditure of a Provisional Sum, as to any possible 'knock-on' effect.

The Quantity Surveyor has the duty to make a fair valuation where the foregoing rules do not cover the situation. He will exclude from his valuation any amounts for direct loss and/or expense and the clause further requires him not to include any direct loss or expense in the valuation of a variation which would be reimbursed elsewhere in the Conditions, i.e. Clauses 26 and 34.

13.6   Where work has to be measured on site for the purposes of valuing a variation, the Contractor is entitled to be present. Remember that it is possible for the measurement to be made from drawings but the entitlement to be present still stands.

13.7   As a cross-check to Clause 3 – Contract Sum, all valuations will be added to or deducted from the Contract Sum.

Scotland

Clause 13 – Variations and Provisional Sums

There are no changes to the above clause for use in Scotland.

However, a minor point should be made. The RICS agreement of a Definition of Prime Cost of Daywork in Scotland is with the Electrical Contractors' Association of Scotland and not with the Electrical Contractors' Association.

Chapter 8

Variation to the works' duration

Clause 25 – Extension of time
Clause 26 – Loss and expense caused by matters materially affecting regular progress of the works

England and Wales

Clause 25 – Extension of time

Although an extension to the time of a contract will not in itself entitle the Contractor to any extra payments, it can decrease the time he may be late in achieving Practical Completion, and so save him from the liability to pay damages referred to in Clause 24 and discussed in a later chapter. It is very important that the Contractor should carefully record all that happens on site and be in a position to request an extension of time, if the circumstances dictate. As can be seen, the period for which the Contractor is in default over non-completion can have a restriction or price adjustment to fluctuations (Clauses 38.4.7 and 40.7 refer).

An extension of time is a further period beyond the original completion date stated in the Appendix, which will then establish a new completion date. An extension will be granted only when the following procedures are accurately and timeously carried out:

Action sequence

Contractor

25.2.1.1   If it becomes reasonably apparent that the progress of the works is being or is likely to be delayed, then the Contractor will give written notice to the Architect with an explanation of what has happened or is likely to happen, and also state which one or more Relevant Events refer.

Note the words 'reasonably apparent'; the Contractor is expected to act responsibly in his supervision and it is in his own interests to forecast likely delay. Also the notice has to be in writing and sent immediately, not at the end of the Contract.

25.2.1.2   A nominated sub-contractor affected by the above notice must receive a copy of the notice.

25.2.2   For all the relevant events referred to in the Contractor's notice, he has to give the Architect, in writing, either at the same time, or as soon as possible,

       25.2.2.1 Particulars of the expected effects,

       25.2.2.2 An estimate of the expected delay beyond the completion date whether or not concurrent with delay from any other relevant event. All this information shall also be given to any nominated sub-contractor who had a copy of the Contractor's original notice.

25.2.3   The Contractor will give the Architect and any affected nominated sub-contractors any further notices, in writing, that he considers necessary or as the Architect requests, in order to up-date the estimate and intimate any changes in the information

It is envisaged here that a somewhat continuous dialogue be established between the Architect and the Contractor to attempt to keep the whole situation in a close focus. It is in the latter's own interests to keep the Architect fully informed, as, in his turn, he has to form an opinion as to the award of an extension, based on the Contractor's evidence.

Architect

25.3.1   If after receiving from the Contractor any notices, particulars and estimates, it is the Architect's opinion that

       25.3.1.1 Any of the delays stated by the Contractor are caused by a relevant event,

       25.3.1.2 The completion of the works is likely to be delayed beyond the completion date,

       then he shall, in writing, give the Contractor an extension of time by fixing a later completion date which he *then* estimates to be fair and reasonable, with a copy to every nominated sub-contractor, even those not notified under Clause 25.2.12.

Notice again that the Architect's obligation in fixing a new completion date is to be fair and reasonable in the estimation of the extension of time, and that this can only be in regard to the cause of the delay. As you will see later, the Contractor has a duty to try to reduce the affects of delay, and his actions may well have some bearing on the Architect's decision.

When the Architect fixes a new completion date, he will state

       25.3.1.3 Which of the relevant events he has taken into account,

       25.3.1.4 The extent, if any, to which he has taken into account any instruction requiring the omission of work as a variation issued since the fixing of the last completion date.

If practicable, the new completion date shall be fixed by the Architect not later than 12 weeks after the Contractor's notice has been received, and if that time does not remain, then not later than the completion date in the Appendix.

This underlines the importance of the Contractor giving not only a written notice but all the particulars and estimates as well, since the Architect cannot, and is not obliged to, give a decision until he has received them all, and then the 12-week period will commence.

25.3.2 Even after the granting of an extension of time, the completion date could be fixed at an earlier date because of a variation for the omission of work being taken into account by the Architect. This is always provided that the Architect considers the new date to be fair and reasonable and that the variation was issued after the date of the last grant of an extension of time. (No new date can be fixed that is earlier than the original date in the Appendix.)

25.3.3 The Architect will make a final review of the progress of the Contract within a 12-week period, starting at the date of Practical Completion, and state in writing to the Contractor (with a copy to every nominated sub-contractor) one of the following three things:

25.3.3.1 Fix a fair and reasonable new completion date taking into account any of the relevant events, whether reviewing a previous decision or otherwise, and whether or not the Relevant Event has been notified by the Contractor in a notice,

OR

25.3.3.2 Fix a fair and reasonable new completion date earlier than that previously fixed because of a variation for the omission of work issued after the date of the last grant of an extension of time,

OR

25.3.3.3 Confirm the last fixed completion date.

The Architect's decision is affected by the Contractor's always using his best endeavours to prevent delay in progress, however caused, and preventing the works from being completed beyond, or further beyond, the completion date (Clause 25.3.4.1 refers).

The Contractor's best endeavours would not be taken to mean working overtime, bringing on new plant, or increasing his labour force significantly, which would all amount to a serious extra expense. Rather it could be an adjustment to his programme in the short term, or possibly a re-allocation of labour.

25.3.4.2 The Contractor has to satisfy the Architect that he has done all that he can reasonably do to proceed with the works.

25.3.5 As mentioned previously, the Architect will notify (in writing) every nominated sub-contractor of each decision in the fixing of a completion date.

25.3.6 As stated before, no new completion date can be fixed which is earlier than that given in the Appendix.

List of Relevant Events

25.4.1   *Force majeure*

This is sometimes quoted as an 'Act of God' but could probably have a wider meaning. It is intended to cover matters of a very exceptional or extreme nature which are outside the control of either party. Care must be taken not to include any of the Perils in Clause 22 under this heading.

25.4.2   Exceptionally adverse weather conditions

This can be difficult to establish; e.g., is the weather 'exceptional' either for this time of year or in its duration? Local weather records over a reasonably long time span would be required to produce what could be reasonably termed 'normal' and then compared with what is claimed to be 'exceptional'. 'Adverse' also has to be considered; frost and rain are not the only problems. A long period of extreme heat could also have an adverse effect on the Contractor's operations.

25.4.3   Loss or damage under Clause 22 Perils

The list of Perils and their exclusions are contained in Clause 1.3. *See* Chapter 2.

25.4.4   Civil commotion, local combination of workmen, strike or lockout affecting

any of the trades employed upon the works,
any trades engaged in preparation,
any trades engaged in manufacture,
any trades engaged in transportation.

Clause 22 Perils covers riot and commotion affecting the works. This clause covers all the building operations before the works are actually built, on or off site and including a possible transport strike.

25.4.5.1   Compliance with Architect's instructions under Clauses

2.3   Discrepancy between documents
13.2   Variations
13.3   Instructions for expenditure of Provisional Sums
23.2   Instruction for postponement
34   Antiquities
35   Nominated Sub-contractors
36   Nominated Suppliers

All Architect's instructions under these clause numbers can give cause for delay, and in every case the Contractor must have well-documented records to enable proof of delay to be established to the Architect's reasonable satisfaction.

25.4.5.2   Opening-up for inspection, or testing materials and goods, including making good (unless the fault can be attributed to the Contractor)

If the work or materials are not up to standard as set out in the Contract Bills, not only does the Contractor have to bear the total cost of rectification, but there is no question of an extension of time over any delay caused by the opening-up, testing and making good (Clause 8.3 refers).

25.4.6   Not having received from the Architect, in due time, necessary instructions, drawings, details or levels for which the Contractor specifically applied in writing. Provided that the date it was requested (bearing the completion date in mind) was neither unreasonably distant from, nor unreasonably close to, the date on which the information was needed

The essence here is that the Contractor must have requested the information and in writing. 'In due time' must allow him the usual margin to order materials, get them delivered and so allow him properly to plan his operations. As to when the request was made, it would be unreasonable to request details of finishings before the work is out of the ground. However, it is not unreasonable to plan fairly well in advance, and thus any request for information would be in good time. Conversely, if the application was too late for the information the Contractor can only have himself to blame and would have no claim for an extension under this clause. Again the Contractor is urged to supervize the Contract properly and record his requests for information and the dates upon which it was received.

25.4.7   Delay on the part of Nominated Sub-contractors and Suppliers which the Contractor has tried to avoid or reduce

This means delay in the completion of Nominated Sub-contract Works which, in turn, causes delay in the progress or completion of the Main Works. Again, well-documented site records with dates are necessary. The Architect will also expect the Contractor to have taken some kind of avoiding action to help the situation.

25.4.8.1   The execution of or failure to execute, work outwith the Contract, either by the Employer direct or by persons engaged by him

Clause 29 defines the type of work and persons engaged to carry it out and this event adequately covers the Contractor against delay on their part.

25.4.8.2   The supply or failure to supply goods which the Employer has agreed to provide

The comment on the previous clause is relevant here.

25.4.9   If after the date of tender the Government or other statutory body passes any edict or law which directly affects the works by

    restricting the supply or use of labour,
    preventing or delaying the supply of goods and materials, and/or
    preventing or delaying the supply of fuel or energy

    all of which are essential to the carrying out of the works.

This would cover, as it suggests, some governmental ruling, such as a shorter working week, fuel rationing or a ban on the importing of certain materials or goods. Note, however, that it must be essential to the carrying out of the works.

25.4.10.1   The Contractor's inability to secure labour, which is not within his control and could not have been reasonably foreseen

The expression 'reasonably foreseen' means in effect that the Contractor should have taken some preliminary steps to make local enquiries, such as whether or not another contract being built at the same time is severely decreasing the local labour pool. The criterion is that the situation is outside the Contractor's control, even though he has made reasonably efforts and possible some additional expenditure.

25.4.10.2   The Contractor's inability to secure such goods and materials, which is not within his control and could not have been reasonably foreseen

Again, the comments above are valid here. Perhaps if the choice of material is restricted to one supplier who is not nominated, then the Contractor's position is easier to establish, but he is expected to have made reasonable efforts.

25.4.11   The carrying out of, or failure to carry out, work by a local authority or statutory undertaker

This event is straightforward and self-explanatory, but consider a failure to supply to the site electric power for tools or plant (temporary supply). Further, if the sewer connection work to be executed by a local authority was later than promised or programmed, would this actually have an effect on the regular progress of the works? Therefore, care has to be taken as to the effect of the delay; it is not an automatic grant of an extension of time.

25.4.12   Failure of the Employer to give in due time

(a) Ingress or egress from the site of the works or part of the works,
(b) Through or over any land, buildings, way or passage next or connected to the site and in the Employer's control,
(c) In accordance with the Contract Bills and/or drawings,
(d) After receipt by the Architect of any required notice to be given by the Contractor,
(e) Failure to give ingress or egress by the Employer as otherwise agreed between the Architect and the Contractor.

It is fairly normal for the Contract Bills or a particular layout drawing to have ingress and egress set out. It is also possible that there is a restriction imposed, maybe one entrance in and another way out. The giving of a notice will be detailed in the Contract Documents and this could involve fencing-off parts of a playground, or a playing field, or re-routing pedestrians for their own safety. Failure by the Employer to allow the Contractor to do these things, in due time, will obviously upset his programme of work and cause delay to the progress of the works. Further, he needs to be instructed about what to do next in consequence of the Employer's default. Thus, the Architect could confirm the failure by instruction which becomes a variation under Clause 13.1.2. A further Relevant Event could then arise, possibly under Clause 25.4.5.1.

Do not confuse this Event with possession of the site in Clause 23. Here ingress and egress are involved although it is difficult to see how one can have possession without getting ingress, so default here could lead to default under Clause 23.

Diagram 1 illustrates the possible sequence of happenings with regard to Clause 25.

Scotland

Clause 25 – Extension of time

There are no changes to this clause for use in Scotland.

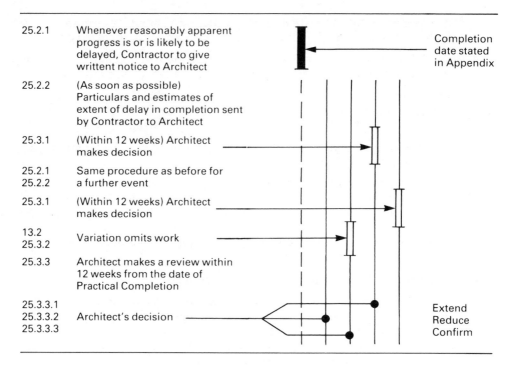

25.2.1	Whenever reasonably apparent progress is or is likely to be delayed, Contractor to give writtent notice to Architect	Completion date stated in Appendix
25.2.2	(As soon as possible) Particulars and estimates of extent of delay in completion sent by Contractor to Architect	
25.3.1	(Within 12 weeks) Architect makes decision	
25.2.1 25.2.2	Same procedure as before for a further event	
25.3.1	(Within 12 weeks) Architect makes decision	
13.2 25.3.2	Variation omits work	
25.3.3	Architect makes a review within 12 weeks from the date of Practical Completion	
25.3.3.1 25.3.3.2 25.3.3.3	Architect's decision	Extend Reduce Confirm

Diagram 1: Clause 25 – Extension of time – possible sequence of happenings

England and Wales

Clause 26 – Loss and expense caused by matters materially affecting the regular progress of the works

This clause in particular provides the vehicle for contractual claims. However, it is not to be regarded as all-embracing, as it is governed by (i) the stipulation that regular progress has been materially affected, and (ii) a fairly narrow list of matters on which such a claim for loss and expense is to be based. It should be further recognized that it is meant for reimbursement only where there are no other provisions in the Contract. It must also be pointed out that the clause's provisions are not final, as Clause 26.6 reserves the Contractor's rights and remedies still to go to arbitration or the Courts.

Although it would appear that Clauses 25 and 26 are directly related, the list of Relevant Events and the list of matters are not the same in their entirety. Because an extension of time is granted it does not follow that there is an automatic claim for loss and expense. A claim for loss and expense could be valid without an extension of time first being granted. The opposite is also true. A contract with late information from the Architect regarding details of the external works, may give rise to a claim for loss and expense but not an extension of time. It is generally considered that the grant of an extension of time makes the validity of a claim for loss and expense very much easier to substantiate.

A comparison of the two lists clarifies the situation.

	Relevant events	Matters
Force majeure	25.4.1	—
Exceptionally adverse weather	25.4.2	—
Loss or damage Clause 22 Perils	25.4.3	—
Civil commotion, strikes, etc.,	25.4.4	—
Compliance with Architect's instructions under		
Clauses 2.3, 13.2, 13.3, 23.2, 34, 35, 36	25.4.5.1	—
under Clause 23.2	—	26.2.5
under Clauses 13.2, 13.3	—	26.2.7
Opening-up work for inspection and testing		
where all is satisfactory	25.4.5.2	26.2.2
Delay in receipt of information	25.4.6	26.2.1
Delay by Nominated Sub-contractors and Suppliers	25.4.7	—
Employer's failure to do work	25.4.8.1	26.2.4.1
Employer's failure to supply materials	25.4.8.2	26.2.4.2
Government restrictions on labour or materials	25.4.9	—
Non-availability of labour	25.4.10.1	—
Non-availability of goods or materials	25.4.10.2	—
Work by statutory undertakers	25.4.11	—
Employer's failure to give ingress or egress	25.4.12	26.2.6
Discrepancies between documents	25.4.5.1	26.2.3
Antiquities	25.4.5.1 refers	although not in the list of matters 34.3.1 refers

26.1   This first sub-clause details the procedures to take place and it is very important that these procedures be followed carefully to protect rights to claim loss and expense. The action procedure is laid out as follows:

Contractor

(1) If in the Contractor's opinion he has, or is likely to have, direct loss and/or expense because the regular progress of the works has been affected materially by one or more of the matters, he will make written application to the Architect (always with the proviso that he could not in his opinion be paid under any other provision of the Contract).

(2) If required by the Architect, the Contractor will furnish him with all relevant information to enable his opinion to be formed as to the effect on regular progress.

(3) If required by the Architect, or the Quantity Surveyor, the Contractor will furnish the details of his claim for loss and/or expense.

From the Contractor's point of view it is better to make up the details of the claim whilst the circumstances are still fresh in everybody's mind, even if the Architect or the Quantity Surveyor do not request details immediately. Again it must be remembered that the

Contractor will write to the Architect as soon as a claim becomes apparent to him and the Architect will do nothing until he has received such a notice. Further, the latter can not ascertain an amount until the Contractor furnishes details of how the loss and/or expense is arrived at. The Contractor could of course have a partial ascertainment of any one claim or claims.

Architect

(1) His opinion has to be formed on whether or not there is any effect on the regular progress of the works due to the matters contained in the Contractors written application.

(2) If his opinion is positive, either he, or the Quantity Surveyor at his request, must ascertain the amount of loss and/or expense.

(3) Any such amounts which are ascertained are to be included in Interim Certificates.

Sub-clause 26.5 directs the Architect to add any ascertained amounts to the Contract Sum from time to time. This could be taken to mean that part of a claim could be agreed and paid for although another part might still be in dispute. No retention will be held on these amounts.

26.3   When the Architect ascertains the amount of loss and/or expense, and he has taken into account a previous award of extension of time, this shall be stated to the Contractor in writing and detailing the relevant events.

26.4   When a nominated sub-contractor makes a written claim to the Contractor for loss and/or expense, it is the latter's clear duty immediately to pass on that claim to the Architect.

The Architect carries out the same procedure as before, with the exception that the notification of extension of time and the pertinent relevant events which he has taken into account in his ascertainment, is sent to the Contractor and a copy sent to the Nominated Sub-contractor.

'Claim' is an emotive word and is not used in the Conditions; it is stated as 'making application for reimbursement'. However, regardless of the actual wording, Clause 26 is an attempt to specify in advance of the events, the rights and remedies of both parties, in certain defined circumstances. Without this clause, these events would be breaches of the Contract.

The Contractor must recognize that he can receive his just reward under the terms of the Contract, but grossly inflated or exaggerated claims without basis in fact or common sense will lead inevitably to dispute and bad feeling. He will be entitled to loss of profit but not to a profit larger than he originally included in the priced Bill of Quantities.

If the Contractor does have loss and/or expense because of delay on the part of Nominated Sub-contractors or Suppliers, his right of recovery is within his Contract with them, NSC/4 or NSC/4a.

The ascertainment of loss and/or expense is in the opinion of the Architect or the Quantity Surveyor and the Contractor could disagree with their opinions. Sub-clause 26.6 reserves the latter's right to start arbitration or court proceedings if a solution is otherwise unobtainable.

Scotland

Clause 26 – Loss and expense caused by matters materially affecting regular progress of the works

There are no changes to this clause for use in Scotland.

Chapter 9

Insurances

Clause 20 – Injury to persons and property and Employer's indemnity

Clause 21 – Insurance against injury to persons and property

Clause 22 – Insurance of the works against Clause 22 Perils

England and Wales

Generally it can be said that insurances are intended to *reinforce* indemnities but in no way take their place. Indemnity means the protection which one party to an agreement affords to the other against some category of loss, or against the claims of a third party. As such, indemnity is very wide in scope but is also unlimited in the potential liability of the one party to the other within that scope. This means that the protection offered will depend upon the financial ability of the Contractor to meet his liability. Insolvency of the Contractor could come about in meeting his liabilities. In view of this, the Contract provides for specific insurance cover to reinforce the indemnity.

The following Table sets out the insurance responsibilities and exceptions, of the two parties, as contained in the three clauses:

| Clause number | Responsibilities for insurance/indemnity | | Exceptions |
	Employer	Contractor	
20.1		Indemnity – injury to persons	Act of neglect by employer or his agents
20.2		Indemnity – injury to property	Any risks under 22B or 22C
21.1.1.1		(a) Personal injury or death during the progress of the works	Act of neglect by Employer or his agents
		(b) Injury to property, real or personal, during the progress of the works	
21.2.1		In joint names – adjoining property (Prov. Sum in Bills)	Clauses 21.2.1.1 to 21.2.1.5
22A.1		In joint names – all Perils (new building)	Exclusions Clause 1.3
22B.1.1	All Perils (new building)		Ditto
22C.1.1	All Perils (existing building and extensions)		Ditto

Clause 20 – Injury to persons and property and Employer's indemnity

20.1   As can be seen from the Table this includes personal injury or death which is the Contractor's liability arising out of, or caused by the carrying out of the works. If injury or death was caused by the alleged negligence of the Employer, the Contractor would have to show proof.

Here any person making a claim must, in law, be fit to do so. A trespasser or a thief would not be legally able to sustain a claim.

In the exception, the clause states 'Employer or his agents'. Presumably these agents would include the Architect, Quantity Surveyor, Consultants, etc.

If the Employer has a claim against him in respect of injury or death, he can either sue you to recover, or become a third party to the litigation. Either way the insurance should be effective, to cover the situation and the indemnity obligations.

20.2   Again, from the Table, this involves injury to property, real or personal, which is the Contractor's liability arising out of or caused by the carrying out of the works. Here the reverse from Clause 20.1 operates and the Employer has to prove the Contractor's negligence, or that of his Sub-contractors. Further, if either Clause 22B or 22C operates, which are the Employer's risk (Perils), the Contractor is not liable for indemnification to the Employer. He is not of course, liable to indemnify against damage caused by the exclusions to the Perils (Clause 21.3).

A claim against the Employer would be handled here in the same manner as in Clause 20.1.

Note that 'property' also includes damage to the works. In all of Clause 20 it must be fully understood that indemnity is unlimited and any limitations in the insurance cover have no effect on the Contractor's indemnity liability.

Clause 21 – Insurance against injury to persons and property

21.1   This clause states that the Contractor will maintain insurance to cover for personal injury or death and damage to property as required by Clause 20 and so honour his liabilities imposed by indemnity. Although the wording states only 'maintain such insurances as are necessary', the Contractor must read into this such a time scale as covers all his periods on site. Possibly it would be as well to maintain his insurance until the date of the Certificate of Making Good Defects.

It is the Contractor's responsibility to ensure that all his Sub-contractors also maintain such insurances. This is mentioned again in later Chapters covering the various contracts between the Contractor and his Domestic and Nominated Sub-contractors.

In respect of personal injury or death, the Contractor has a statutory liability to conform with the Employer's Liability (Compulsory Insurance) Act 1969 and any amendments thereto.

The insurance cover for all of Clause 21.1 is stated as a minimum amount in the Appendix, but the Contractor can, of course, cover for a greater sum. He would be well advised to consult an insurance broker in regard to all his insurance commitments.

The Architect has the power to require the Contractor, or any of his Sub-contractors, to produce proof that the necessary insurances are properly maintained and ask for inspection of the policies or premium receipts. Should the Contractor, or any of his Sub-contractors, default in maintaining the proper insurances, the Employer can himself insure against any risk and recover the cost from monies due to the Contractor, or as a debt.

21.2   This clause states that the Contractor will maintain insurance in his own and the Employer's names jointly in respect of damage to property other than the works. The liability of the Employer is restricted here to a list of reasons, set out below, causing damage to other property during the carrying out of the works:

   (a) Collapse,
   (b) Subsidence,
   (c) Vibration,
   (d) Weakening or removal of support,
   (e) Lowering of ground water.

There are stated exceptions to the causes of damage where all liability is relieved, as under:

   (a) Negligence, omission or default of the Contractor,
   (b) Errors or ommissions in design,
   (c) Matters which could not have been reasonably foreseen,
   (d) Sole risk of the Employer under Clause 22B or 22C,
   (e) Nuclear risk, war risk or sonic booms.

The Contractor is required to take out this insurance only when there is a Provisional Sum included in the Contract Bills and an Instruction is issued as to its expenditure. The amount of cover is stated in the Bills, the insurer approved by the Architect and the policy and premium receipts given to him for safe keeping.

The Contractor will recover the amounts expended in maintaining the insurance, by their being added to the Contract Sum. This would then be in the form of a variation which would be valued according to the rules, with the Provisional Sum being deducted. The Contractor would be entitled to add all other costs he incurs over and above the premiums, as this would be the total amount for maintaining the insurance.

As before, should the Contractor default in the maintenance of the insurance, then the Employer himself may effect the insurance and the amount of the premiums will not be included in any adjustment of the Contract Sum. It might be worth noticing that only the amount paid, or payable in respect of premiums, is excluded from any Contract Sum adjustment. This means that the Employer would not recover any other costs over and above the premium payments, in effecting the insurance.

21.3 This is simply a standard list of insurance exclusions, all of which relieve the Contractor of liability for indemnification and insurance. The list of exclusions can be found in Clause 1.3 under the definition of Clause 22 Perils.

Clause 22 – Insurance of the works against Clause 22 Perils

22 This clause is split into three distinct parts, namely A, B, and C, of which only one can operate in a Contract, and the operative clause will be that which is not struck out of the Appendix.

From the Table the three options are:

22A In joint names for a new building if Contractor is to insure against the Perils.

22B For a new building if Employer is to insure against the Perils.

22C For alterations or extensions to an existing building where the Employer insures against the Perils.

A footnote to this clause indicates that in some cases it might not be possible to obtain insurance against some of the Perils. In this circumstance, the Architect should be informed, and dependent upon the action taken and arrangements made between the parties, the appropriate clause would be adjusted.

22A This clause refers to a new building where the Contractor has the responsibility for the insurance against the Perils and it is taken out in the names of the Employer and the Contractor jointly.

It is of paramount importance to understand that the insurance cover is for the full reinstatement value together with any professional fees. The insurance must be maintained until the issue of the Certificate of Practical Completion, which means making due allowance for higher building costs, inflation, or any other financial burdens there may be to put the building back to its original state after loss or damage, by any one or more of the Perils. It is further stated that the cover extends to materials or goods on site, but not to the Contractor's tempory site huts or buildings. Professional fees are usually stated in the Appendix as a percentage; thus the monies the Contractor will receive will be only the insurance money less the amount for professional fees. With this amount he has to complete the reinstatement as he is required to under this clause.

Again the Architect has the right to approve the insurers, and the Contractor must deposit with him the policy and premium receipts. If the Contractor defaults in this insurance, the Employer can insure and deduct the amounts of the premiums from monies due to the Contractor, or if none, recover as a debt.

If the Contractor has a normal Contractor's all-risks policy which would be adequate to cover the amount required for loss or damage by Perils, then further insurance in the joint names is not required, provided that the Employer's interest is endorsed on the policy. Check that the amount is enough to cover the amount for professional fees. The Architect has the right, in this case, to inspect the policy and premium receipts, but it is not necessary to deposit the policy or premium receipts with him.

Any default by the Contractor in insurance in this case, will be dealt with in the same manner as that of a policy in the joint names.

The Contractor's main contractual obligation is to carry out and complete the works. Accordingly, as soon as an insurance claim is accepted by the insurance company, he has to start restoring damaged work or replacing materials on site. Removal of debris is also part of his responsibility, the cost of which is part of the full reinstatement value.

As it is the Employer who receives the insurance monies, payment to the Contractor will be by installments in the normal Interim Certificates and he will not receive any other payment for the restoration work other than that recovered from the insurers.

Note that the Contractor's liability for reinstatement would cease if determination took place under Clause 28.1.3.2.

22B   This clause refers to a new building where the Employer has the responsibility for the insurance against the Perils.

As mentioned before under Clause 20.2, because the Employer is to bear the risk of insurance against the Perils, the Contractor is then relieved of this liability for indemnity. It could be argued further that in terms of the Perils his contractual liability to complete the works ceases and becomes the Employer's, but it is specifically written into the Clause that the Contractor will make the restoration and carry out and complete the works.

The Employer will maintain the insurance and the Contractor has the right to inspect the last premium receipt of renewal. Note, however, that the Contractor still needs to insure his own temporary huts, buildings, plant and equipment.

If the Employer defaults, the Contractor can insure in the Employer's name, and by producing proof of a premium receipt have that amount added to the Contract Sum.

Upon discovering loss or damage to the works or unfixed materials or goods, the Contractor will immediately notify the Employer and the Architect, stating the extent, nature and location. It would also be beneficial here to mention the Peril or Perils causing the loss or damage.

As any loss or damage is not included in payments to the Contractor at the time, it is for the Architect to certify the value of the work executed prior to the loss or damage.

The Contractor has to resinstate the damage and go on to complete the works, including removing debris. This work will be treated as a variation, and valuation and payment will be made in accordance with the rules.

Note that the only exceptions to the insurance are those as set out in Clause 1.3; therefore, the Employer, by effecting this insurance, accepts the liability for the Perils when the loss or damage could be caused by negligence either of the Contractor or of his Sub-contractors.

22C This clause refers to alterations and extensions to an existing building where the Employer has the responsibility for the insurance against the Perils.

The insurance here also covers the contents of the buildings, but again the Contractor has to insure his own temporary huts, buildings, plant and equipment.

Again the Employer will maintain the insurance and the Contractor has the right to inspect the last premium receipt of renewal. Should the Employer default, the Contractor can insure in the Employer's name, and by producing proof of a premium receipt have that amount added to the Contract Sum. Because the contents also are included in the insurance, in this case the Contractor has right of entry and inspection to enable him to make a survey and inventory of the existing structures and their contents.

Upon discovering loss or damage to the works or unfixed materials and goods, the Contractor will immediately notify the Employer and the Architect, stating the extent, nature and location. As before, indicating the Perils would be helpful.

As any loss or damage to the works is not included in payments to the Contractor at the time; it is again for the Architect to certify the value of the work executed prior to the loss or damage.

Reinstatement under this clause is not automatic since it is possible for the loss to be total and bear no relationship to the original Contract. Therefore, either the Employer or the Contractor can, within 28 days of the occurrence, institute determination proceedings, by notice by registered post or recorded delivery. Within seven days, (note, not longer) of the receipt of such notice by either party, then either party may give the other a written request to agree to the appointment of an Arbitrator to decide whether or not such a determination is just and equitable.

Following on either a notice of determination or a request for arbitration, one of two things will happen:

(1) The Arbitrator upholds the notice of determination and Clause 28.2 operates, with the exception of Clause 28.2.2.6 (*see* Chapter 15).

(2) Where no determination has taken place within 28 days or the Arbitrator has decided against the notice, the Contractor will proceed to reinstate the damage and complete the works. The Architect can issue an instruction for the removal and disposal of debris. The work of reinstatement and removal of debris (if instructed) will be treated as a variation, and valuation and payment made in accordance with the rules.

Scotland

Clause 20 – Injury to persons and property and Employer's indemnity

Clause 21 – Insurance against injury to persons and property

Clause 22 – Insurance of the works against Clause 22 Perils

There are no changes to the above clauses for use in Scotland.

However, the Scottish Supplement does set out the three alternatives, (22A, 22B and 22C) and defines their uses. It further reinforces the fact that only one alternative is to be chosen, with the others being struck out. These alternatives are found in Appendix II of the Scottish Building Contract.

It is also worth noting that in Scotland, it is the Arbiter and not the Arbitrator who is appointed for an arbitration.

Chapter 10

Payments

Clause 16 – Materials and goods unfixed or off-site
Clause 30 – Certificates and payments
Clause 31 – Finance (No 2) Act 1975 – statutory tax deduction scheme England and Wales

Clause 16 – Materials and goods unfixed or off-site

Unfixed materials and goods, once delivered to site, can not be removed without the Architect's written consent. Here the Architect has to be reasonable, since it is possible that cutting or finishing of some material might best be done in the Contractor's own workshop.

When the value of unfixed materials and goods has been properly paid in accordance with Clause 30, they become the property of the Employer. Regardless of ownership, while the materials and goods are on site the Contractor is responsible for loss and damage, unless Clause either 22B or 22C is applicable to the Contract. Transfer of ownership or title of the materials and goods to the Employer may not be automatic just because they have been paid for by the Employer. The Contractor must have secure title in the materials and goods to enable a transfer of ownership to take place. In other words, the Contractor must have a contract of supply, or proof of payment to the supplier.

Unfixed materials and goods which are intended to be fixed into the works but which are stored off-site, and have had their value properly paid in accordance with Clause 30, also became the property of the Employer. These materials and goods cannot be moved from their storage position (unless to the site) without the Architect's permission. Again, regardless of ownership, the Contractor is responsible for loss and damage, and storing, handling and insurance costs. As soon as delivery to site has taken place, the materials and goods are in the same category as those unfixed, except that having been paid for they are already the Employer's property.

Clause 30 – Certificates and payments

This is quite a long and complicated clause, but a clear understanding of how it operates is necessary for the Contractor to ensure a continuous cash-flow for his operations.

In the practical sense, it is worth noting that the Nominated Quantity Surveyor will make up the valuation for interim and final payments, which could be with or without the Contractor's co-operation. Dependent upon the size of Contract and his organization, it is quite normal for his Surveyor and the Nominated Quantity Surveyor to prepare valuations together.

The valuation of a payment is passed to the Architect and it is he who issues the payment certificate. It is the Employer who pays out the monies. The Contractor will find that at this stage the Employer has rights of deduction, other than retention, from monies due to him, and these rights will be discussed in detail as they occur in the various sub-clauses.

Two diagrams are given here in an effort to simplify the clauses and pull them together as a whole. The first, *Diagram 2*, shows the payment sequences throughout the Contract, with the minimum and maximum time intervals between certificates, and the second, *Diagram 3*, shows the interaction between the Design Team, Contractor and his Nominated Sub-contractors and Suppliers.

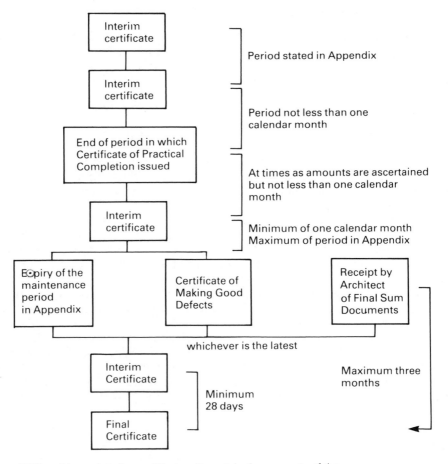

NB Penultimate interim certificate will contain the amounts of the finalized sub-contract sums for all Nominated Sub-contracts.

Diagram 2: Payment sequence (Sub-clauses 30.1.3, 38.7 and 30.8)

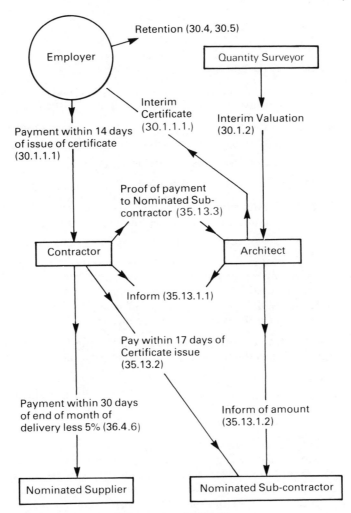

Diagram 3: Interaction of Clauses 30, 35 and 36 – Interim Certificates

30.1.1.1   The Architect will issue certificates of payment to the Contractor at time intervals. These interim periods (set out in the Appendix) are usually one month, and the Contractor is entitled to payment within 14 days from the date of issue of each interim certificate.

Non-payment by the Employer within 14 days gives the Contractor the right to begin determination proceedings under Clause 28.1.1.

The Employer has rights of deduction from monies due to the Contractor as certified by the Architect, in accordance with the following clauses:

4.1.2   Architect's Instructions not complied with; cost of having work done by others employed by the Employer.

21.1.3   Cost of insurance premiums where the Contractor has defaulted.

21.4.2   As above.

22A.2   As above.

22A.3   As above.

24.2.1   Liquidated and Ascertained Damages (usually applicable only to Penultimate and Final Certificates).

27.4.2.2   Payments to Suppliers or Sub-contractors due to determination by the Employer.

31   Statutory tax deduction scheme.

35.13.5.3   Payments to Nominated Sub-contractors.

35.24.6   Additional costs due to nomination of further Sub-contractors after the Sub-contractor's determination.

—   Obligations under the VAT Agreement.

30.1.1.2   The Employer cannot deduct any amounts relating to the above clauses from monies held as retention, unless it is released retention in a certificate. Released retention becomes monies due.

There is also a reference to Clause 35.13.5.4.2 which places a restriction on the monies to be deducted.

30.1.1.3   The Contractor must be informed in writing by the Employer of the reasons for deductions being made from monies due to him.

30.1.2   Valuations of interim payments will be made by the Quantity Surveyor. Any fluctuations under Clause 40, the formula adjustment method, require that an interim valuation shall be made before the issue of each Interim Certificate (Clause 40.2).

30.1.3   This clause is covered graphically under Payment Sequence. It should be noted that the Expiry of the Maintenance Period and the Certificate of Making Good Defects are considered in tandem, using whichever date is the latest. Also the three months' maximum period for the Final Certificate starts from the latest of three dates, namely, Expiry of the Maintenance Period, Certificate of Making Good Defects and receipt by the Architect or Surveyor of all documents to enable the computation of the adjusted Contract Sum.

30.2   The amount to be paid in any Interim Certificate, subject to any agreement as to stage payments, will be

(a) The Gross Valuation
   less
(b) Retention
(c) Previous payments stated to be due in previous interim certificates

All monies as set out in the following clauses and valued up to and including a date not more than seven days before the date of issue

Note that 'seven days before the date of issue' is qualified as 'not more than', the obvious inference being, that less than seven days is acceptable. Ideally, the valuation and date of issue of the certificate should be as close together as possible.

Computation of gross valuation

Additions

NB It will be apparent that as this is the exhaustive list not all these clauses will necessarily be applicable to any or all interim valuations.

30.2.1.1   Total value of the work properly carried out by the Contractor, including any work carried out under Clause 13.5 (Variations) together with any applicable adjustment of the value under Clause 40 (Fluctuations).

30.2.1.2   Total value of unfixed materials and goods delivered to or adjacent to the works, provided that this value will only be for materials and goods reasonably, properly, and not prematurely delivered, and adequately protected against weather and other casualty (e.g. damage).

30.2.1.3   Total value of unfixed materials and goods off site at the Architect's discretion for inclusion in an Interim Certificate.

30.2.1.4   Amounts calculated for Nominated Sub-contracts.

30.2.2.5   Under Clauses 21.4.1 and 21.4.2 of Sub-contract NSC/4 or NSC/4a.

30.2.1.5   Contractor's profit on Nominated Sub-contract amounts last referred to, calculated at the rates in the Contract Bills, or when nomination arises from an instuction as to the expenditure of a Provisional Sum, at rates related thereto, or if none, at reasonable rates.

30.2.2.1   Amounts changing the Contract Sum under Clause 3, namely

    6.2   Payment of fees and charges.

    7   Architect's Instructions regarding levels and setting out the works.

    8.3   Instructions to open up for inspection and tests.

    9.2   Instructions with regard to royalties and patient rights.

    17.2   Making good of defects, shrinkages or o. er faults.

    17.3   Instructions regarding defects, shrinkages o other faults.

    21.2.3   Amounts expanded by the Contractor to maintain insurance.

    22B   As above.

    22C   As above.

30.2.2.2   Amounts ascertained under Clauses

    26.1   Direct loss and/or expense.

    34.3   Direct loss and/or expense due to antiquities.

30.2.2.3   Amounts to which Clause 35.17 refers, final payment to Nominated Sub-contractors.

30.2.2.4   Any amounts payable under Clause 38 or 39 Fluctuations.

Omissions

30.2.3.1   Any amount allowable to the Employer under Clause 38 or 39 Fluctuations.

30.2.3.2   Amounts referred to in Clause 21.4.3 of Sub-contract NSC/4 or NSC/4a – Discounts.

30.4.1   The retention which the Employer may deduct and retain is such percentage of the total amount included under Clauses numbered

30.2.1.1
30.2.1.2
30.2.1.3   } *See* gross valuation additions
30.2.1.4
30.2.1.5

—   Previous payments stated to be due in previous Interim Certificates.

Retention

Clauses 30.4 and 30.5 deal exclusively with the rules relating to retention, which are summarized in the following, together with a tabulated check-list showing the amounts on which the various stages of retention operate.

The Retention percentage is stated as 5 per cent unless stated otherwise in the Appendix. It is further stated that all contracts estimated to be of a Contract Sum of £500 000 or over, the Retention should be 3 per cent.

All retention deducted will be specified under the two headings of Contractor's Retention and Nominated Sub-contract Retention. A statement specifying the amounts of each of these Retentions deducted from an Interim Certificate must accompany the certificate and be issued to the Employer, the Contractor and each Nominated Sub-contractor concerned.

Notwithstanding the Employer's trusteeship of the retention monies, he does not need to invest it, but can do so without being accountable for any interest gained. Both Contractor and Nominated Sub-contractors are here safeguarded against the possibility of bankruptcy of the Employer, as the Retention trust money cannot be used to pay out any of the Employer's creditors.

When retention is released in an Interim Certificate, it then becomes 'monies due', and accordingly the Employer can exercise his right to deduct (cross-check Clause 30.1.1.2). It is stated again here that if this is done, a statement detailing such deductions must be given to the Contractor and relevant Nominated Sub-contractors.

In the Private Edition either the Contractor or a nominated sub-contractor, either separately or together, or at different times, can request the Employer, at the date of payment of each certificate, to place the retention in a separate banking account, named to identify the amount as the retention, held by the Employer on trust and for whom. This request must be granted by the Employer, and certified to the Architect (with a copy to the Contractor) that such amount has been so placed. The Employer will be entitled to any interest and not have to account for same. Although this will secure the retention for the Contractors, it would not normally be requested unless the financial stability of the Employer was in doubt. The bank account would state the Employer's name and that the account was in trust for (the Contractor's name) and also the name of the Contract. For most Architects and Contractors the bank statement would suffice as certification of the account.

Retention check list

Full retention 5% or 3% or other agreed per cent	Half retention	No retention
Work which has not reached Practical Completion (Clauses 17.1, 18.1.2, 35.16)	Work which has reached Practical Completion (Clauses 17.1, 18.1.2, 35.16)	Any amounts included under Clauses 30.2.2.1 to 30.2.2.5 inclusive (*see* Gross Valuation Additions)
Materials and goods on or off site (Clauses 30.2.1.2, 30.2.1.3, 30.2.1.4)	AND FOR WHICH NONE OF THE UNDER-MENTIONED HAVE BEEN ISSUED	
	Certificate of Completion of Making Good Defects (Clause 17.4)	
	OR	
	Certificate of Making Good of any defects, shrinkages or other faults (Clause 18.1.3)	
	OR	
	Interim Certificate for final payment of Nominated Sub-contractors (Clause 35.17)	

By the operation of the above the Employer will release to the Contractor one-half of the retention upon payment of the next Interim Certificate after Practical Completion. After the expiration of the Defects Liability Period named in the Appendix, or after the issue of the Certificate of Completion of Making Good Defects (whichever is the later), the balance of the retention will be released. This refers to the whole or part of the works as is applicable.

Final account

Before, or within a reasonable time after Practical Completion of the work, the Contractor will send to the Architect or, if requested, to the Quantity Surveyor, all documents necessary for the adjustment of the Contract Sum, including those relating to the accounts of Nominated Sub-contractors and Nominated Suppliers. It is fair to say that in the majority of contracts the preparation of the final account is an on-going thing, and documentation will have been provided in the case of any sub-contractor whose work is finished before the Certificate of Practical Completion.

Provided that the Contractor has supplied the necessary documents to the Quantity Surveyor, the latter will prepare a statement of all the final valuations, including Nominated Sub-contractors, which will be done during the period of final measurement and valuation, as stated in the Appendix. The Architect will send a copy to the Contractor and relevant extracts to the Nominated Sub-contractors.

It is perhaps interesting to note that, according to the Conditions, the period of Final Measurement and Valuation is six months from the date of Practical Completion, and this could be different if a start date for the period was entered in the Appendix. This is another case where careful scrutiny of the Appendix is important. Also the Quantity Surveyor is not obliged to produce the final valuations until the Contractor has produced all the necessary documents, and delay on his part will hold up the issue of the Final Certificate. This responsibility is still the Contractor's in respect of both his Domestic and Nominated Sub-contractors.

Clause 30.6.2 details the format of the adjustment of the Contract Sum, which is reproduced here in a summarized version.

The Contract Sum adjustment or amount of Final Account

	Deduct			*Add*	
30.6.2.1	(a) All prime cost sums (b) All amounts of Nominated Sub-contractors and Suppliers (c) Any Contractor's profit thereon		**30.6.2.6**	The total amounts of Nominated Sub-contract Sums or Tender Sums as finally adjusted	
30.6.2.2	(a) All provisional sums (b) All work described as 'provisional' in the Contract Bills		**30.6.2.7**	Amount of the Tender Sum accepted by the Employer from the Main Contractor under Clause 35.2	
30.6.2.3	(a) Amount of omissions of any Variation Order (b) Amount in the Contract Bills for any other work which has its conditions changed due to a Variation Order (Clause 13.5.5)		**30.6.2.8**	Final amounts for goods and materials of Nominated Suppliers (note 5% cash discount)	
			30.6.2.9	Profit on accounts of Nominated Sub-contractors and Suppliers	
30.6.2.4	Any amount allowable to the Employer, due under Clauses 38, 39 and 40 (Fluctuations)		**30.6.2.10**	Reimbursement of monies due because of payments made under Clause:	
30.6.2.5	Any other amount which is required to be deducted from the Contract sum			**6.2** Fees and charges **7** Levels and setting out **8.3** Opening up for inspection **9.2** Royalties and patent rights **17.2** Defects and shrinkages **17.3** Defects and shrinkages **21.2.3** Insurance premiums	
			30.6.2.11	Amount of valuation of any variation including work referred to in Clause 13.5.5	
			30.6.2.12	Amount of work in expenditure of provisional sums, or work described as provisional in the Contract Bills	
			30.6.2.13	Any amount ascertained under **26.1** Loss and expense **34.3** Antiquities	
			30.6.2.14	Amounts paid by the Contractor under Clauses 22B or 22C insurance premiums (Private Edition only)	
			30.6.2.15	Amounts under Clauses 38, 39 and 40 (Fluctuations)	
			30.6.2.16	Any other amount required to be added to the Contract Sum	

The Contractor will receive a copy of the computation of the adjusted Contract Sum prior to the issue of Final Certificate. This gives him at least the full period of 14 days allowed for starting arbitration proceedings, beginning on the issue of the Final Certificate.

Clauses 30.7 and 30.8 are recorded graphically in the Payment Sequence. It can be seen that the 28-day period between the Interim and Final Certificates gives the Architect ample time to check the Contractor's payments to Nominated Sub-contractors. Of course, if proof of payment is not made within this period then the Final Certificate will be delayed.

The Final Certificate will state the sum of the amounts already stated as due in Interim Certificates, as well as the Adjusted Contract Sum. The difference between these two amounts is the amount of the Final Certificate due either to the Contractor, or by him to the Employer, as the case may be. This difference is subject to any deductions authorized by the Conditions and without prejudice to the Contractor's rights in respect of any Interim Certificates not paid by the Employer. Fourteen days after the date of issue of the Final Certificate, the amount or balance becomes a debt payable by either party.

Except in the case of Arbitration Clauses 30.9.2 and 30.9.3, or in the case of fraud, the Final Certificate means:

Conclusive evidence that the quality of materials and the standard of workmanship, are to the reasonable satisfaction of the Architect, if such is his remit.
Conclusive evidence that all adjustments, i.e. amounts to be added or deducted, have been effected in regard to the Contract Sum, excepting

any accidental omission or addition in the final computation,
any arithmetical error in the final computation.

These exceptions apart, the Final Certificate is conclusive evidence of all other computations.

If arbitration or other proceedings have been started by either party *before* issue of the Final Certificate, it will then be only conclusive evidence as above, after the earliest of either

Proceedings have been concluded, whereupon the Final Certificate will be subject to any award from such proceedings,

OR

A period of 12 months in which neither party has taken any proceedings further, whereupon the Final Certificate will be subject to any terms agreed in partial settlement.

If any arbitration or other proceedings have been started by either party within 14 days *after* the Final Certificate's issue, it will then only be conclusive evidence, save only in respect to which these proceedings relate.

Other than the Final Certificate, no other Architect's Certificate is conclusive evidence that the works, materials and goods are in accordance with the Contract. Note that the Final Certificate can be limited in its extent.

Off-site materials and goods

This Sub-clause 30.3 has been deliberately taken out of context to enable the Scottish Contractor to follow Clause 30 more easily as a whole, as this sub-clause does not operate North of the Border.

The Architect, at his discretion, can include in any Interim Certificate the value of materials and goods off site, always provided that the following appropriate criteria are satisfied:

The materials are intended for inclusion in the works.

Nothing remains to be done to the materials to complete them, prior to their inclusion in the works.

The materials have been set aside at the premises where they have been manufactured, assembled, or stored and are clearly marked individually, or in sets, by some form of reference, or code, to establish their identity. To inform the Employer where they are stored on the premises of the Contractor and in any other case, the person to whose order they are held and their destination as the works.

Where the materials were ordered from a supplier by the Contractor, or any sub-contractor, the Contract for supply must be in writing and provide that ownership will pass unconditionally to the Contractor, or Sub-contractor, not later than their completion for inclusion in the works and labelling to establish their identity.

Where the materials were ordered from a supplier by a sub-contractor, the Sub-contract must be in writing and provide that ownership passing to the sub-contractor will immediately pass to the Contractor.

Where the materials were manufactured or assembled by a sub-contractor, the Sub-contract must be in writing and provide that ownership will pass unconditionally to the Contractor not later than completion for inclusion in the works and labelling to establish their identity.

The materials are in accordance with this Contract.

The Contractor provides the Architect with reasonable proof of ownership and that all the previous criteria have been complied with.

The Contractor provides the Architect with reasonable proof that the materials are insured for their full value under a policy protecting the interests of the Employer and the Contractor, in respect of the Perils, for the period between transference of ownership to the Contractor and their delivery to the works.

If should be noted that payment for off-site materials is at the Architect's discretion and he will wish all the relevant appropriate criteria to be satisfied before inclusion of the amount in any Interim Certificate.

Clause 31 – Finance (No 2) Act 1975 – statutory tax deduction scheme

This clause states, if you like, the rules to provide an arrangement contractually which are necessary to operate the tax deduction scheme in the construction industry. All payments by 'Contractors' to 'Sub-contractors' have tax deducted at source, unless the 'Sub-contractor' can prove exemption by holding a current Inland Revenue tax exemption certificate. For a total understanding, the phamphlet *Construction Industry Tax Deduction Scheme* is published by the Inland Revenue.

31.1   Here is a glossary of definitions of terms.

31.2   At the date of tender, the Appendix has to state whether or not the Employer is a Contractor in the terms of the Act. If he is *not* a Contractor then the remainder of Clause 31 does not operate. Although it must seem odd that some Employers could also be Contractors, remember that a client could well have a construction business interest, or be a local authority with a 'direct works' construction department, both of which cases would make them contractors under the Act, as well as employers.

Even though the Appendix states the Employer is not to be a contractor, he could become one before the date of the Final Certificate. In that case, he must inform the Contractor and the remainder of Clause 31 would operate.

31.3   Not later than 21 days before the first Interim Certificate is due, or after the coming into operation of the previous paragraph (Clause 31.2.2), the Contractor shall prove his tax certificate is valid and current, or inform the Employer in writing (with a copy to the Architect) that he is not entitled to be paid without tax deduction.

Should the Employer not be satisfied with the proof provided above, he will, within 14 days of the proof being submitted, write to the Contractor to inform him that he will be making a tax deduction, and give his reasons for doing so. The Employer must then also follow through the procedures laid down in Clause 31.6.1.

31.4   This sub-clause sets out the Contractor's action in regard to tax certificates, for the following alternatives:

(a) Obtaining a first-time tax certificate with regard to this contract.
(b) Tax certificate to be out of date before final payment.
(c) Cancellation of current tax certificate.

31.5   The Employer will promptly send to the Inland Revenue all vouchers which the Contractor as a sub-contractor gives him.

31.6   If, in the Employer's opinion, a tax deduction has to be made, he will tell the Contractor in writing and ask him to state, not later than seven days before each future payment is due (or within ten days of his notification, if that is later), the amount which is the direct cost to the Contractor or any other person, of materials used or to be used. This then leaves the labour element.

When the Contractor does as requested, he will indemnify the Employer for any incorrect statement (there could be loss and expense caused to the Employer).

If the Contractor does not do as requested, the Employer will make his own fair estimate of the amount.

31.7   Calculation errors will be corrected by the Employer, by repayment or deduction from payments, always subject to any statutory obligation on the employer not to make a correction

31.8   Complying with Clause 31 takes precedence over other Conditions.

31.9   Article 5 will apply to any dispute between the parties over the operation of Clause 31 unless the Act or the regulations, or any other rule, or order, states otherwise.

Scotland

Clause 16 – Materials and goods unfixed or off site

The Scottish Supplement adds to the end of Sub-clause 16.1 the words 'and for any materials and/or goods purchased prior to their delivery to the site under the separate Contract referred to in Clause 30.3 hereof'. Sub-clause 16.2 is wholly deleted.

Both these amendments are to bring Clause 16 into line with the Scottish alternative to Sub-clause 30.3 where goods and materials off site have to be the subject of a separate contract of purchase.

Clause 30 – Certificates and payments

The Scottish Supplement, reproduced opposite, wholly deletes Clauses 30.2.1.3 and 30.3. Clause 30.3 is substituted and a new Clause 30.11 added.

Clause 30.3 is so worded as to allow the Employer, with the Architect's advice as to the expediency, to enter into a Contract of Purchase for off-site materials and goods, with the Contractor or any Sub-contractor. This entails the completion of two separate forms, one for each case, to form a Contract, both of which are appended and commented upon in the next few pages.

Clause 30.11 enables the Employer to secure Liquidated and Ascertained Damages by adjusting the balance due to, or by, the Contractor, in the Final Certificate.

Scottish Supplement

Delete		Add
Clause 30.2.1.3 wholly (materials and goods off site)		Document references would be NSC/4/Scot or NSC/4a/Scot
NB Under Clause 30.2.2.5 Under Clause 30.2.3.2	**30.3**	If the Architect/Supervising Officer is of the opinion that it is expedient to do so, the Employer may enter into a separate contract for the purchase from the Contractor or any sub-contractor of any materials and/or goods prior to their delivery to the site, which the Contractor is under obligation to supply in terms of this Contract and upon such contracts being entered into the purchase of the said materials and/or goods shall be excluded altogether from this Contract and the Contract Sum shall be adjusted accordingly. Provided that when the Employer enters into a separate contract with any sub-contractor
Clause 30.3 wholly (materials and goods off site)		
	30.3.1	he shall do so only with the consent of the Contractor, which consent shall not be unreasonably withheld, and
	30.3.2	payment by the Employer to the Sub-contractor for any of the said materials and/or goods shall no way affect any cash discount or other emolument to which the Contractor may be entitled and which shall be paid by the Employer to the Contractor.
	30.11	Nothing in Clauses 30.6.2 or 30.9.2 shall prevent the Employer from deducting or adding Liquidate and Ascertained Damages in accordance with Clause 24 hereof or from any sum due by him to the Contractor or by the Contractor to the Employer as the case may be under the Final Certificate.

SBCC

CONTRACT

between

.... Employer

and

.... Contractor

relative to the purchase of materials and/or goods
required for use at

The constituent bodies of the Scottish Building
Contract Committee are: —

Royal Incorporation of Architects in Scotland

Scottish Building Employers Federation

Scottish Branch of the Royal Institution of Chartered
    Surveyors

Convention of Scottish Local Authorities

Federation of Specialists and Sub-Contractors
    (Scottish Board)

Committee of Associations of Specialist Engineering
    Contractors (Scottish Branch)

Association of Consulting Engineers
    (Scottish Group)

Copyright of the S.B.C.C., 39 Castle Street, Edinburgh

February 1980

Contract

between

(hereinafter referred to as "the Employer")

and

(hereinafter referred to as "the Contractor")

_____

CONSIDERING THAT the Employer and the Contractor have entered into a Building Contract dated whereby the Contractor is carrying out certain works for the Employer at (hereinafter referred to as "the Works").

FURTHER CONSIDERING that the undernoted materials and/or goods are required for use in the Works, and that they are the property of the Contractor at the date hereof.

THEREFORE the Employer has agreed to purchase and hereby purchases from the Contractor the following materials and/or goods:

Type and quantity of materials and/or goods	Where set aside	Identifying mark or number

on the following terms and conditions: —

1   The said materials and/or goods shall be used by the Contractor in the execution of the Works.

2   The purchase price payable by the Employer shall be £ payable in two instalments viz: —

    2.1   £   payable on the date hereof*

    2.2   £   payable on the later of the following events

        2.2.1   the expiry of the defects liability period for the Works, or in the case of a Sectional Completion Contract, the expiry of the defects liability period for the Section in which the said materials and/or goods are employed, or

        2.2.2   the date of issue by the Architect of the Certificate of making good defects for the Works or Section as the case may be.

*_The amount of the first instalment should be 5% (or the rate of retention if different) less than the total purchase price payable by the Employer and the second the balance of that sum._

3   All clauses of the said Building Contract except the provisions thereof for payment shall apply to the said materials and/or goods as if they were materials and/or goods delivered to placed on or adjacent to the Works and intended therefor, and included in an Interim Certificate in respect of which the Contractor had received payment.

4   The property in the said materials and/or goods shall pass to the Employer on payment of the sum stated in 2.1 above.

5   The Contractor shall not, except for use in the Works, remove or cause or permit the same to be removed from the premises where they have been manufactured or assembled or are stored.

6   The Contractor shall be responsible for any loss thereof or damage thereto and for arranging and paying for storage, handling and insurance of the same which insurance shall cover the Clause 22 perils as defined in the said Building Contract as well as cover for theft, malicious damage and loss or damage in transit, and shall be subject to the conditions referred to in Clause 22A of the said Building Contract.

IN WITNESS WHEREOF these presents are executed at   on the
day of   19   before these witnesses subscribing: —

.... witness

.... address

.... ocpupation

                                        .... Employer
.... witness

.... address

.... occupation

.... witness

.... address

.... occupation

                                        .... Contractor
.... witness

.... address

.... occupation

Scottish Building Contract Committee

Instructions for using the Contract of Purchase attached when the materials and/or goods are the property of the Main Contractor

Clause 30.3 of the Standard Form of Building Contract gives the Architect a discretionary power to authorise payment for off site materials and/or goods which does not conform with Scots law. This clause and the connected Clause 16.2 are therefore deleted in the Scottish Building Contract.

There will be occasions in Scotland when an Architect will want to exercise such a discretion and, if so, he must therefore see that the Employer and Contractor enter into the attached form of Contract of Purchase for the particular materials and/or goods. The Architect's attention is particularly drawn to the fact that the payment under the Contract of Purchase is specifically excluded from certification under the Building Contract and he should therefore make a corresponding reduction in the Contract Sum. He must also satisfy himself before advising the Employer to purchase the said materials and/or goods that the following conditions are satisfied:

(a) Such materials and/or goods are intended for incorporation in the Works;

(b) Nothing remains to be done to such materials and/or goods to complete the same up to the point of their incorporation in the Works;

(c) Such materials and/or goods have been and are set apart at the premises where they have been manufactured or assembled or are stored, and have been clearly and visibly marked, individually or in sets either by letters or figures or by reference to a pre-determined code, so as to identify:
   (i) (a) The Employer where they are stored on premises of the Contractor or
       (b) Where they are not stored on premises of the Contractor, the person to whose order they are held, and
   (ii) in either case, their destination as being the Works.

(d) Where such materials and/or goods were ordered from a supplier by the Contractor or by a Sub-Contractor or were manufactured or assembled by a Sub-Contractor, the contract for their supply, manufacture or assembly is in writing and expressly provides that the property therein shall pass unconditionally to the Contractor or Sub-Contractor not later than the happening of the events set out in paragraphs (b) and (c) above.

(e) Where such materials and/or goods were ordered from a supplier by a Sub-Contractor the sub-contract between the Contractor and the Sub-Contractor is in writing and expressly provides that on the property passing to the Sub-Contractor it shall immediately pass to the Contractor.

(f) The materials and/or goods are in accordance with the Building Contract.

(g) The Contractor provides the Architect with reasonable proof that the property in such materials and/or goods belongs to him and that the conditions set out in paragraphs (a) to (f) have been complied with.

Care must be taken that a receipt is obtained from the Contractor for the first part of the price of the materials and/or goods when payment is made as this will complete the documentary proof that the property in the materials and/or goods has passed to the Employer.

Explanatory notes

The SBCC Contract between the Employer and Contractor for the purchase of materials and/or goods off site.

This blue form is fairly straightforward in approach and simple to follow. The First page requires the names and addresses, the date of formation, or date of acceptance letter of the Main Contract and its title or name.

Following below is the necessary description of the type and quantity of the goods, where they are stored and their identification. It is always best to have the identifying number or name physically on the goods, such as by stencilling, not a loose, tied or stapled tag which could come off. Remember that each batch of goods, when ready, requires a separate Contract of Purchase, the criteria being that the goods are completed, stored and properly marked.

Clause 1 requires the checking of the quantity of goods for the Contract, in case of any excess over goods actually required.

Clause 2 establishes that the retention is the same as that of the Main Contract and sets out the two-stage payment.

Clause 3 second page allows the operation of all the Main Contract conditions, with the exception of payment.

Clause 4 lays down the time of passage of ownership and is reinforced by the last paragraph of the instruction sheet.

Clause 5 reiterates the Contractor's responsibility in Clause 16 of the Main Contract.

Clause 6 covers the insurance responsibility, not only for Perils but also for theft, damage and loss or damage in transit. Note that storage, handling and insurance charges are all the Contractors responsibility.

Any doubts which may arise in executing this formal Contract should be taken to legal advice before signature.

The instructions for using the Contract of Purchase are clearly set out and intended to be strictly adhered to.

SBCC

CONTRACT

between

.... Employer

and

.... Contractor

and

.... Sub-Contractor

relative to the purchase of materials and/or goods
required for use at

The constituent bodies of the Scottish Building
Contract Committee are: —

Royal Incorporation of Architects in Scotland

Scottish Building Employers Federation

Scottish Branch of the Royal Institution of Chartered
Surveyors

Convention of Scottish Local Authorities

Federation of Specialists and Sub-Contractors
(Scottish Board)

Committee of Associations of Specialist Engineering
Contractors (Scottish Branch)

Association of Consulting Engineers
(Scottish Group)

Copyright of the S.B.C.C., 39 Castle Street, Edinburgh

February 1980

Contract

between

(hereinafter referred to as "the Employer")
and

(hereinafter referred to as "the Contractor")
and

(hereinafter referred to as "the Sub-Contractor")

CONSIDERING THAT the Employer and the Contractor have entered into a Building Contract dated   whereby the Contractor is carrying is carrying out certain works for the Employer at
(hereinafter referred to as "the Works") and that in connection therewith the Contractor has entered into a Building Sub-Contract with the Sub-Contractor dated

FURTHER CONSIDERING that the undernoted materials and/or goods are required for use in the Works and that the said materials and/or goods are the property of the Sub-Contractor at the date hereof.

THEREFORE the Employer with the consent of the Contractor has agreed to purchase and hereby purchases from the Sub-Contractor the following materials and/or goods:

Type and quantity of materials and/or goods	Where set aside	Identifying mark or number

on the following terms and conditions:

1   The said materials and/or goods shall be used by the Sub-Contractor in the execution of the Works.

2   The purchase price payable by the Employer to the Sub-Contractor with the consent of the Contractor shall be £   payable in two instalments viz: —

  2.1   £   payable at the date hereof*

  2.2   £   payable on the later of the following events

    2.2.1   the expiry of the defects liability period for the Works, or in the case of a Sectional Completion Contract, the expiry of the defects liability period for the Section in which the said materials and/or goods are employed, or

    2.2.2   the date of issue by the Architect of the Certificate of making good defects for the Works or Section as the case may be.

*The amount of the first instalment should be 5% (or the rate of retention if different) less than the total purchase price payable by the Employer and the second the balance of that sum.

3   All clauses of the said Building Contract and Building Sub-Contract except the provisions thereof for payment shall apply to the said materials and/or goods as if they were materials and/or goods delivered to placed on or adjacent to the Works and intended therefor and included in an Interim Cetificate in respect of which the Contractor had received payment and as if the Sub-Contractor was also the Contractor under the Building Contract.

4   The property in the said materials and/or goods shall pass to the Employer on payment of the sum stated in 2.1 above.

5   Neither the Contractor nor the Sub-Contractor shall, except for use in the Works, remove or cause or permit the same to be removed from the premises where they have been manufactured or assembled or are stored.

6   The Contractor shall be responsible for any loss thereof or damage thereto and for arranging and paying for storage, handling, insurance of the same which insurance shall cover the Clause 22 perils as defined in the said Building Contract as well as cover for theft, malicious damage and loss or damage in transit, and shall be subject to the conditions referred to in Clause 22A of the said Building Contract:

IN WITNESS WHEREOF these presents are executed at   on the
day of   19   before these witnesses subscribing:

.... witness
.... address
.... occupation
                              .... Employer
.... witness
.... address
.... occupation

.... witness
.... address
.... occupation
                              .... Contractor
.... witness
.... address
.... occupation

.... witness
.... address
.... occupation
                              .... Sub-Contractor
.... witness
.... address
.... occupation

Scottish Building Contract Committee

Instructions for using the Contract of Purchase attached where the material and/or goods are the property of a Sub-Contractor

Clause 30.3 of the Standard Form of Building Contract gives the Architect a discretionary power to authorise payment for off-site materials and/or goods. This discretionary power is extended to sub-contracts by Clause 21.2.3 of the Nominated Sub-Contracts NSC/4 and NSC/4a. None of these clauses conform with Scots Law and are therefore deleted in the Scottish Building Contract and the Scottish Building Sub-Contracts NSC/4/Scot and NSC/4a/Scot respectively.

There will be occasions in Scotland when an Architect will want to exercise such a discretion in favour of a Sub-Contractor and if so he must therefore see that the Employer, the Contractor and Sub-Contractor enter into the attached form of Contract of Purchase for the particular materials and/or goods. The attention of the Architect and the Contractor is particularly drawn to the fact that the agreed sum due under the Contract of Purchase in so far as it is contained in the Contract Sum is specifically excluded from certification under the Building Contract and corresponding reductions should be made in the Contract and Sub-Contract Sums. The Architect must also satisfy himself before advising the Employer to purchase the said materials and/or goods that the following conditions are satisfied:

(a) Such materials and/or goods are intended for inclusion in the Works.

(b) Nothing remains to be done to such materials and/or goods to complete the same up to the point of their incorporation in the Works.

(c) Such materials and/or goods have been and are set apart at the premises where they have been manufactured or assembled or are stored and have been clearly and visibly marked individually or in sets either by letters or figures or by reference to a predetermined code so as to identify:
  (i) (a) the Employer where they are stored on premises of the Sub-Contractor or
      (b) where they are not stored on the premises of the Sub-Contractor, the person to whose order they are held, and
  (ii) in either case, their destination as being the Works.

(d) Where such materials and/or goods were ordered from a supplier by the Contractor or by a Sub-Contractor or were manufactured or assembled by a Sub-Contractor, the contract for their supply, manufacture or assembly is in writing and expressly provides that the property therein shall pass unconditionally to the Contractor or Sub-Contractor not later than the happening of the events set out in paragraphs (b) and (c) above.

(e) Where such materials and/or goods were ordered from a supplier by a Sub-Contractor the Sub-Contract between the Contractor and the Sub-Contractor is in writing and expressly provides that on the property passing to the Sub-Contractor it shall immediately pass to the Contractor.

(f) The materials and/or goods are in accordance with the Building Contract and Sub-Contract.

(g) The Sub-Contractor provides the Architect with reasonable proof that the property in such materials and/or goods belongs to him and that the conditions set out in paragraphs (a) to (f) above have been complied with.

Care must be taken that a receipt is obtained from the Sub-Contractor for the first part of the price of the materials and/or goods when payment is made as this will complete the documentary proof that the property in the materials and/or goods has passed to the Employer.

Explanatory notes

The SBCC Contract between the Employer, the Contractor and a Sub-contractor for the purchase of materials and/or goods off site

This orange form is a contract which this time is between the three parties as above. The form can be used for either a domestic or a nominated Sub-contractor.

Again, there are instructions attached to the end of the form which should be carefully followed, and the criteria must be satisfied.

Note that if the Architect's opinion favours the Contract then both the Employer and the Contractor must also agree. Presumeably the Sub-contractor's agreement is automatic, as he will, in the first place, be the party requesting payment for the goods or materials.

The Sub-contract must be in writing and provide expressly for ownership passing to the Contractor at the same time as it passes to the Sub-contractor.

The last paragraph of the instructions states that a receipt is obtained from the Sub-contractor for the first payment from the Employer. As the receipt goes directly to the Employer, it thus provides proof of ownership passing to the Employer.

Note, however, that Clause 6 of the Contract still makes the Contractor bear the responsibility for storage, handling, insurance against Perils, and theft, malicious damage and loss or damage in transit.

With the exception of the above notes, the two contracts are very similar.

Clause 31 – Finance (No 2) Act 1975 – Statutory tax deduction scheme

There are no changes to the above clause for use in Scotland, beyond those of definition in Clause 1.

Chapter 11

Others employed on the site

Clause 19 – Assignment and Sub-contracts

Clause 29 – Works by Employer, or persons employed or engaged by Employer

England and Wales

Clause 19 – Assignment and Sub-contracts

19.1   Neither the Employer nor the Contractor can, without the written consent of the other, assign to a third party his rights, duties, or obligations.

19.2   As the Contractor will probably not have the organisation to carry out all the work himself, he can sub-let parts of the works to other contractors. These other contractors then become his Domestic Sub-contractors with whom he will have a contract but they have no direct link with the Employer. The Clause lays down the condition that the Contractor must have written permission to sub-let, but most tender documents have a list of proposed sub-contractors to complete. Unless the Employer or the Architect makes some definite objection to any firm on the list, then the signing of the Contract would be automatic approval.

    The contract entered into by the Contractor with each of his Domestic Sub-contractors is the Domestic Sub-contract DOM/1 Articles of Agreement, which is governed by its own set of Sub-contract Conditions. The Articles of Agreement and Sub-contract Conditions are the subject of a companion book in this series.

    It is of course, quite proper for any other form of sub-contract to be entered into, as DOM/1 is only the JCT 80 version.

19.3   This particular clause has, unfortunately, been used by some clients as a substitute for the nomination procedure, which it is not, nor is it intended that it be used for such purpose. Any person named under this clause becomes a Domestic and not a Nominated Sub-contractor. As will be seen by examining this sub-clause in detail, the Architect is unlikely to reserve selection of a sub-contractor whom the Contractor would not select for himself.

A list of sub-contractors may be set out in the Bills of Quantities, or attached to them, for the Contractor to select one to carry out work which has been measured or described in the Contract Bills and priced by the Contractor. The following rules govern the procedures:

(a) The list must comprise of a minimum of three persons.
(b) Prior to any binding sub-contract agreement, the list can be added to by the Employer or the Contractor, with mutual consent (any additions must be initialled by both parties).
(c) Prior to any binding sub-contract agreement, if the list of persons able and willing to do the work is less than three, then:
    (i) The Employer and the Contractor will, by agreement, add further names to bring the list up to its minimum three names (again the additions to be initialled by both parties).

    OR

    (ii) The Contractor can carry out the work or sub-let it to a Domestic Sub-contractor.
(d) Under the rules, any person selected becomes a Domestic Sub-contractor.

As can be seen from the rules above, it is possible that the Architect could, at the finish, have a list of three persons, or someone else sub-let by the Contractor, all of whom were outwith his original selection. The Architect has to agree additional names, but he can withhold consent for some good reason. The Contractor can always add names to the list and these will probably be more favourable to you than would the others. The end result may well be that the Architect, does not have the reserve of selection as he would appear to have at the first reading of the clause. This is a very cogent reason for the clause not being used as a substitute for nomination.

If the Contractor has any reason to object to sub-letting to anyone on the named list, he would be prudent to let that objection be known before submitting his tender.

It must always be remembered that all persons on the list, besides being willing to carry out the work, must also be properly experienced and capable of carrying out the type of work required.

It is conditional that employment of the Domestic Sub-contractor under this clause will determine at the same time as determination of the Contractor's own employment.

The Contractor should also note that failure on his part to comply with Clause 19 is a ground for determination by the Employer under Clause 27.1.4.

Note that Nominated Sub-contractors are mentioned here as having separate procedures. Subject to Clause 35.2 (Nominated Sub-contractors his own tender), the Contractor is not required to carry out any work specifically reserved for Nominated Sub-contractors.

Clause 29 – Works by Employer, or persons employed or engaged by Employer

This clause permits work to be carried out under direct contract with the Employer, and it is needed in order to secure a right of entry to the works for these persons, as the Contractor has possession of the site.

29.1   The Contractor must be given information in the Bills of Quantities with regard to the work to be done by other persons. This could be a description of the works, or more detailed information with regard to the time on site, at which stage of the works the persons will start and finish, or any other specific information relative to the Contractor's programme. Such detailed information would be required if he has to furnish a Master Programme. When information is given in the Contract Bills, he must permit the execution of the work.

29.2   Where the Contract Bills *do not* provide the information as above, the Employer may arrange with the Contractor for the execution of the work, with his consent, which is not to be unreasonably withheld. This must apply to additional work and does not give the Architect licence to substitute others to do work properly in the Contract.

29.3   The Contractor is not responsible under Clause 20 for indemnity to the Employer for persons employed or engaged by him. That is, the Contractor has no liability for personal injury or death because of any act or neglect of persons employed or engaged by the Employer under this clause.

It is worth pointing out that several cross-references to other clauses occur because of this clause, namely:

Clause 25.4.8.1 Delay qualifying for extension of time; this clause is a Relevant Event.
Clause 26.2.4.1 Claim for direct loss and expense; this clause is one of the Matters.
Clause 28.1.3.6 Delay because of Clause 29 is a reason, if the whole works are suspended for the period in the Appendix, for determination by the Contractor.

Scotland

Clause 19 – Assignment and Sub-contracts

Clause 29 – Works by Employer, or persons employed or engaged by Employer

No changes to these clauses for use in Scotland.

Chapter 12

Nominated Sub-contractors

England and Wales (Scotland, p.194)

Clause 35 – Nominated Sub-contractors

This clause is probably the most complicated for all parties concerned to understand and operate. It sets out the rules as to when and how the Architect can nominate a sub-contractor, either to supply and fix, or to execute work.

The definitive description of a Nominated Sub-contractor is: such person whose final selection and approval for the supply and fixing of any materials and goods, or the execution of any work, has been reserved to the Architect. Such reservation can be obtained either by the use of a Prime Cost Sum, or by the naming of a sole contractor in the Contract Documents, in order that the Contractor is therefore bound to enter into a sub-contract with those persons.

Nomination is brought about by either 'Basic' or 'Alternative' methods, each of which requires its own set of documents. The Architect's choice as to which method to use does not need explanation here, as in the end result the Contractor will follow the set pattern of form-filling applicable to either the Basic or the Alternative Method.

Nomination as above and defined can then arise under Clauses 35.1.1 to 35.1.4.

It is worth noting that the Architect's right to nominate under Clause 35.1.3 is by Instruction requiring a variation and is always restricted to additional work which is similar to work in the Contract for which the Architect has the right to nominate (by Prime Cost Sum or naming a sole sub-contractor).

Clause 35.1.4 allows for measured work within the Bills or for work unlike any other nominated work, still to be the subject of nomenation, with the Contractor's agreement to the Architect, which the Contractor will not withhold unreasonably.

Section B.9.1 of the Standard Method of Measurement would require a Prime Cost Sum for every nomination to take place, and accordingly this is over-ridden by Clause 35.1.

35.2   This Clause is included to allow the Contractor to tender for work which is to be the subject of a nominated sub-contract. The Appendix has an entry which he can fill in stating his desire to tender for the named work. The provisions are made, however, that in his normal course of business he carries out that type of work and that the Architect is prepared to accept a tender from him.

The Architect still reserves his right to reject the lowest or any tender, so the Contractor has no right to the work even if he submits the lowest tender. However, if his tender is accepted, he cannot sub-let the work to a domestic sub-contractor without the Architect's permission.

Where the intention is to nominate a sub-contractor for an item included in an Architect's Instruction (possibly the expenditure of a Provisional Sum), then FOR THIS CLAUSE it will be deemed to have been included in the Contract Bills and an entry set out in the Appendix. Thus this effectively gives the Contractor the opportunity to tender.

The Clause further states that it will apply to work in the Bills covered by a Provisional Sum for which the Architect wishes to make a nomination. Should the Contractor's tender be accepted, the work will be valued by the Quantity Surveyor as though it were a variation under Clause 13.

As only this Clause 35.2 applies to work for which the Contractor's tender can be accepted, technically he would not then be a nominated sub-contractor.

35.3   This clause sets out the list of documents necessary to conclude a nomination by either the Basic or Alternative Methods. The list is included in that of the JCT puplished documents at the beginning of this book and are further referred to when each method of nomination is discussed in detail.

Procedure for nomination – general

35.4   The Contractor is not bound to enter into a sub-contract with a proposed nominated sub-contractor if he has a reasonable objection. He must make his objection known as soon as possible and not later than the stage at which he sends Tender NSC/1 to the Architect. *See* Stage 7 in *Diagram 4* (p.122).

35.5   It is understood that all nominations will be made by the Basic Method using Tender NSC/1 and Agreement NSC/2, unless it is specifically stated to the contrary that the Alternative Method (Clauses 35.11 and 35.12) is to be used.

The statement indicating that the Alternative Method is to be used must be clearly set out when nomination is being made, by one of these means:

(a) Naming a sole sub-contractor,
(b) Prime Cost Sum in the Bills,
(c) Prime Cost Sum in an instruction regarding the expenditure of a Provisional Sum, or
(d) Prime Cost Sum in an instruction requiring a variation.

The statement must also contain the information as to whether or not the Employer and Sub-contractor will enter into Agreement NSC/2a.

Clause 35.5.2 allows the Architect the opportunity to change from one method to another by an instruction which will be treated as a variation, but only up to a certain stage in each procedure.

The limiting stages are not later than:

Basic Method   – Stage 6, *Diagram 4* (*see* Clause 35.7.1)
Alternative Method – Stage 3, *Diagram 8* (*see* Clause 35.11)

The exception to this is, if either Clause 35.23 or Clause 35.24 operates (nomination does not proceed, or re-nomination is necessary), when a change could be implemented.

Nomination procedure – Basic Method

35.6   Only persons who have tendered on Tender NSC/1 and entered into Agreement NSC/2 may be nominated under this Method. If Clauses 35.11 and 35.12 operate then this is the Alternative Method.

Diagram 4 shows the complete procedure from start to finish, all of which is covered in Clauses 35.7.1 to 35.10.2 of the Conditions.

Diagram 5 shows the procedure if the proposed sub-contractor withdraws (Clause 35.9).

Diagram 6 shows the procedure if the Main Contractor makes reasonable objection (Clause 35.4.2).

Diagram 7 shows the procedure where the Contractor and the proposed sub-contractor are unable to agree within ten working days. *See* Clause 35.8.

Immediately following the diagrams, on pages 126–137, are the necessary forms, indicating where the various parties have to provide information and/or signatures, to affect a nomination, together with explanatory notes.

The timing of the giving of information and/or signatures in the various forms is important, and with this in view, the various stage numbers and names of parties, to accord with *Diagram 4*, are also entered in the forms.

122

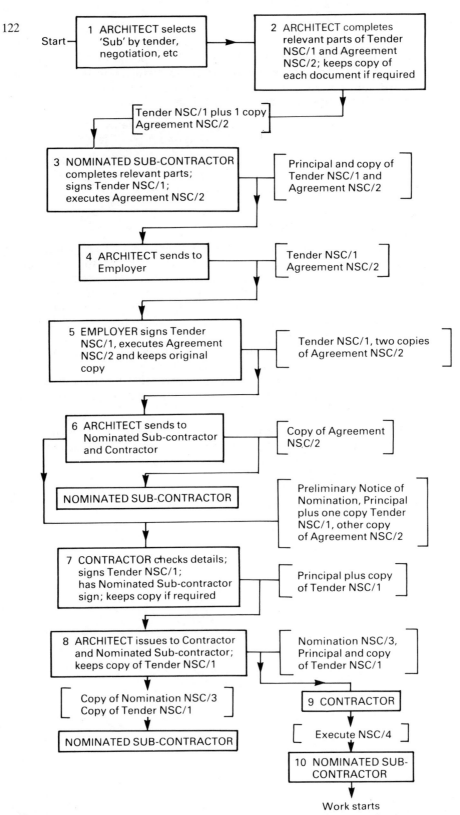

1 ARCHITECT selects 'Sub' by tender, negotiation, etc

Start

2 ARCHITECT completes relevant parts of Tender NSC/1 and Agreement NSC/2; keeps copy of each document if required

Tender NSC/1 plus 1 copy
Agreement NSC/2

3 NOMINATED SUB-CONTRACTOR completes relevant parts; signs Tender NSC/1; executes Agreement NSC/2

Principal and copy of Tender NSC/1 and Agreement NSC/2

4 ARCHITECT sends to Employer

Tender NSC/1
Agreement NSC/2

5 EMPLOYER signs Tender NSC/1, executes Agreement NSC/2 and keeps original copy

Tender NSC/1, two copies of Agreement NSC/2

6 ARCHITECT sends to Nominated Sub-contractor and Contractor

Copy of Agreement NSC/2

NOMINATED SUB-CONTRACTOR

Preliminary Notice of Nomination, Principal plus one copy Tender NSC/1, other copy of Agreement NSC/2

7 CONTRACTOR checks details; signs Tender NSC/1; has Nominated Sub-contractor sign; keeps copy if required

Principal plus copy of Tender NSC/1

8 ARCHITECT issues to Contractor and Nominated Sub-contractor; keeps copy of Tender NSC/1

Nomination NSC/3, Principal and copy of Tender NSC/1

Copy of Nomination NSC/3
Copy of Tender NSC/1

9 CONTRACTOR

NOMINATED SUB-CONTRACTOR

Execute NSC/4

10 NOMINATED SUB-CONTRACTOR

Work starts

Diagram 4: Nomination of a Sub-contractor using the Basic Method

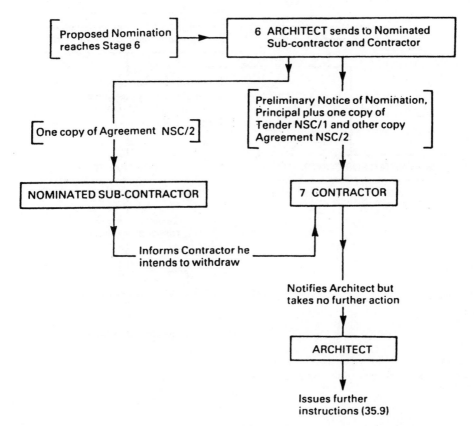

Diagram 5: Nomination of a Sub-contractor using the Basic Method – Proposed Sub-contractor withdraws

124

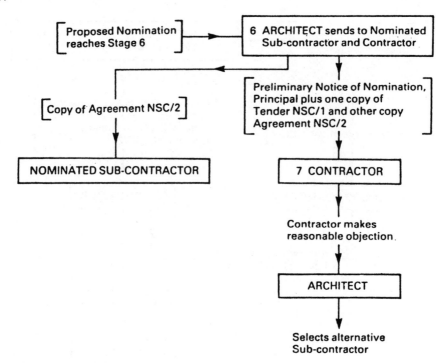

Diagram 6: Nomination of a Sub-contractor using the Basic Method – Contractor makes reasonable objection

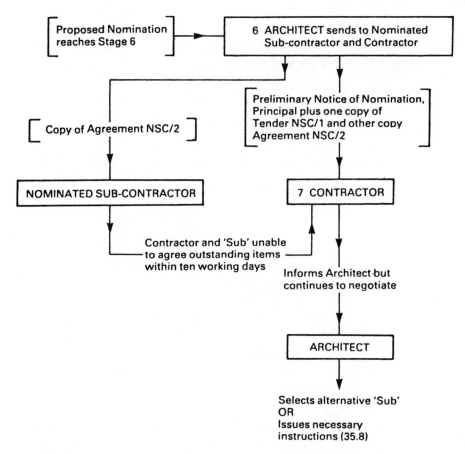

Diagram 7: Nomination of a Sub-contractor using the Basic Method – Contractor and Proposed
Sub-contractor unable to agree within ten working days

126

JCT

JCT Standard Form of Nominated Sub-Contract Tender and Agreement

See "Notes on the Completion of Tender NSC/1" on page 2.

Main Contract Works:[a]

Job reference:

ARCHITECT - STAGE 2
ER STAGE 1

Location:

Sub-Contract Works:

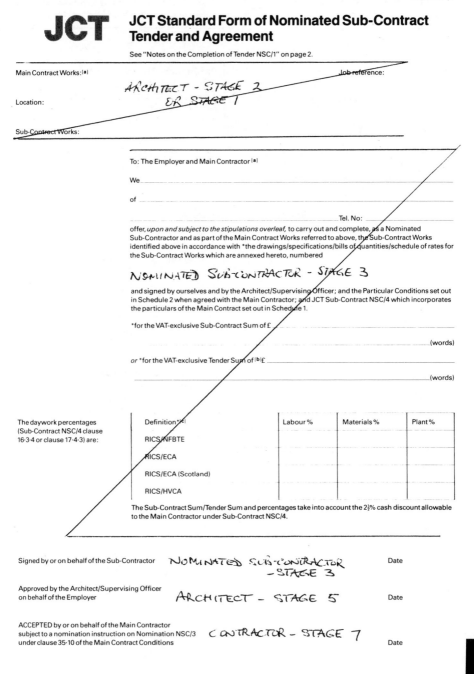

To: The Employer and Main Contractor [a]

We _____

of _____

_____ Tel. No: _____

offer, *upon and subject to the stipulations overleaf*, to carry out and complete, as a Nominated Sub-Contractor and as part of the Main Contract Works referred to above, the Sub-Contract Works identified above in accordance with *the drawings/specifications/bills of quantities/schedule of rates for the Sub-Contract Works which are annexed hereto, numbered

NOMINATED SUB-CONTRACTOR - STAGE 3

and signed by ourselves and by the Architect/Supervising Officer; and the Particular Conditions set out in Schedule 2 when agreed with the Main Contractor; and JCT Sub-Contract NSC/4 which incorporates the particulars of the Main Contract set out in Schedule 1.

*for the VAT-exclusive Sub-Contract Sum of £ _____

_____ (words)

or *for the VAT-exclusive Tender Sum of [b]£ _____

_____ (words)

The daywork percentages (Sub-Contract NSC/4 clause 16·3·4 or clause 17·4·3) are:	Definition*[c]	Labour %	Materials %	Plant %
	RICS/NFBTE			
	RICS/ECA			
	RICS/ECA (Scotland)			
	RICS/HVCA			

The Sub-Contract Sum/Tender Sum and percentages take into account the 2½% cash discount allowable to the Main Contractor under Sub-Contract NSC/4.

Signed by or on behalf of the Sub-Contractor   NOMINATED SUB-CONTRACTOR - STAGE 3   Date

Approved by the Architect/Supervising Officer on behalf of the Employer   ARCHITECT - STAGE 5   Date

ACCEPTED by or on behalf of the Main Contractor subject to a nomination instruction on Nomination NSC/3 under clause 35·10 of the Main Contract Conditions   CONTRACTOR - STAGE 7   Date

© 1980 RIBA Publications Ltd   *Delete as applicable

Tender NSC/1

Stipulations

1. Only when this Tender is signed on Page 1 on behalf of the Employer as 'approved' and the Employer has signed or sealed (as applicable) the Agreement NSC/2 do we agree to be bound by that Agreement as signed by or sealed by or on behalf of ourselves.

2. If the identity of the Main Contractor is not known at the date of our signature on page 1 we reserve the right within 14 days of written notification by the Employer of such identity to withdraw this Tender and the Agreement NSC/2 notwithstanding any approval of this Tender by signature on page 1 on behalf of the Employer.

3. We reserve the right to withdraw this Tender if we are unable to agree with the Main Contractor on the terms of Schedule 2 of this Tender (Particular Conditions).

4. Without prejudice to the reservations in 2 and 3 above this Tender is withdrawn if the nomination instruction (Nomination NSC/3) is not issued by the Architect/Supervising Officer (with a copy to ourselves) under the Main Contract Conditions clause 35·10 within [d]   of the date of our signature on page 1 or such other, later, date as may be notified in writing by ourselves to the Architect/Supervising Officer.

5. Any withdrawal under 2, 3 and 4 above shall be at no charge to the Employer except for any amounts that may be due under Agreement NSC/2.

NOMINATED SUB-CONTRACTOR
— STAGE 3

Notes

[a] For names addresses and telephone numbers of Employer, Main Contractor, Architect/Supervising Officer and Quantity Surveyor see Schedule 1.

[b] Alternative for use where the Sub-Contract Works are to be completely re-measured and valued.

[c] If more than one Definition will be relevant set out percentage additions applicable to each such Definition. There are four Definitions which may be identified here; those agreed between the Royal Institution of Chartered Surveyors and the National Federation of Building Trades Employers; the Royal Institution and the Electrical Contractors Association; the Royal Institution and the Electrical Contractors Association of Scotland and the Royal Institution and the Heating and Ventilating Contractors Association.

[d] Sub-Contractor to insert acceptance period.

[e] See Schedule 2, Item 11.

[f] Delete the two alternatives not applicable. Insert 35·7 or 36·8 percentage.

[g] Insert the same description as in the Main Contract Articles of Agreement.

[h] Delete editions/Supplement as applicable. Insert date of revision.

[i] Insert office address as appropriate.

[j] If a later Completion Date has been fixed under clause 25 this should also be stated here.

[k] Standard Form Local Authorities WITH Quantities edition only.

[l] This information, unless included in the Sub-Contract Specification or Bills of Quantities should be given by repeating here or attaching a copy of the relevant section of the Preliminaries Bill of the Main Contract Bills (or of the main Contract Specification).

[m] Sub-Contractor to complete this Appendix and attach further sheet if necessary.

[n] Date to be inserted by the Architect/Supervising Officer.

[o] To be deleted by the Architect if fuels not to be included. See NSC/4 clause 35·2·1 and clause 36·3·1.

[p] Only applicable where the Main Contract is let on the Standard Form of Building Contract, Local Authorities Edition, WITH Quantities.

[q] If both specialist engineering formulae apply to the Sub-Contract the percentages for use with each formula should be inserted and clearly identified.

[r] The weightings for sprinkler installations may be inserted where different weightings are required.

[s] To be completed by the Architect/Supervising Officer.

[t] The Sub-Contractor will set out here (or on an attached sheet if necessary) details of the carrying out of the Sub-Contract Works as a preliminary indication to the Architect and Contractor. In regard to periods indicate when each is to start. Adaptation will be needed where a phased completion is required.

[u] Not including period required by Architect for approval.

[v] The Contractor to complete in agreement with the Sub-Contractor details of the programme for carrying out the Sub-Contract Works (which must include the subjects set out in 1A and 1B) and to be inserted here or on an attached sheet initialled by the Contractor and Sub-Contractor. The details at 1A and 1B (and in any sheet attached thereto) must then be deleted and the deletion initialled by the Contractor and the Sub-Contractor.

[w] Attention is called to the Order of Works, if any, stated in Schedule 1, Item 11.

[x] The Sub-Contractor will set out here or on an attached sheet as a preliminary indication to the Architect and Contractor details of the attendances which he requires under the headings (a) to (g) which have been extracted from SMM, 6th Edn., B.9.3. These attendances to be supplied at no cost to the Sub-Contractor.

[y] The Contractor to complete and set out in agreement with the Sub-Contractor any alterations to any of the detail of the attendances set out under the headings (a) to (g) in 3-A.

[z] The Contractor to complete in agreement with the Sub-Contractor. This item must include any limits of indemnity which are required in respect of insurances to be taken out by the Sub-Contractor. See NSC/4, clause 7.

[aa] Any special conditions or agreements affecting the employment of labour which the Sub-Contractor wishes to raise should be inserted here.

[bb] The Contractor to complete in agreement with the Sub-Contractor. The details of 5·A. must then be deleted and the deletion initialled by the parties.

[cc] See Sub-Contract NSC/4, clause 24. The Contractor to complete by inserting the names and addresses after agreement with the Sub-Contractor.

[dd] The Contractor to ensure completion as appropriate after agreement with the Sub-Contractor. See Note [i] in Sub-Contract NSC/4, clause 20A on "self-vouchering".

[ee] The Contractor to delete in agreement with the Sub-Contractor.

[ff] The Sub-Contractor will set out here or on an attached sheet any matters he wishes to agree with the Contractor.

[gg] The Contractor to complete in agreement with the Sub-Contractor. The details of 9·A. must then be deleted and the deletion initialled by the parties.

128

Schedule 1: Particulars of Main Contract and Sub-Contract

Names and addresses of:

Employer: Tel. No:

†Architect/Supervising
Officer: Tel. No:

Quantity Surveyor: Tel. No:

Main Contractor: Tel. No:

1. Sub-Contract Conditions: Sub-Contract NSC/4, appropriate to the Standard Form of Building Contract edition identified in item 5
                                                 of this Schedule, unamended: 19   edition (revised.............................). To be executed forthwith
                                                 after Architect/Supervising Officer's nomination on Nomination NSC/3.[e]

2. Sub-Contract   NSC/4 [f] Clause 35 (see also Appendix A)
    Fluctuations: or Clause 36 (see also Appendix A)
                                              35·7 or 36·8......................%
                                        or Clause 37 (see also Appendix B)

3. Main Contract Appendix   Where relevant will apply to the Sub-Contract unless otherwise specifically stated here. The entry
    and entries therein: relating to clause 37 of the Main Contract Conditions is for information of the Sub-Contractor only.
    (see item 10 pages 4 and 5)

ARCHITECT - STAGE 2

4. Main Contract Works:[g]

5. Form of Main Contract   Standard Form of Building Contract, 1980 edition.
    Conditions:
                                        Local Authorities/Private edition/WITH/WITH APPROXIMATE/WITHOUT
                                        Quantities (revised...........................)[h]

                                        Sectional Completion Supplement

6. Inspection of Main   The unpriced *Bills of Quantities/Bills of Approximate Quantities/Specification (which incorporate the
    Contract: general conditions and preliminaries of the Main Contract) and the Contract Drawings may be inspected
                                        by appointment at: [i]

*Delete as applicable

†Note: The expression 'Supervising Officer' is applicable where   Schedules and in Agreement NSC/2. Where the person who will
the nomination instruction will be issued under the Local   issue the nomination instruction is entitled to the use of the name
Authorities Edition of the Standard Form of Building Contract and   'Architect' the expression 'Supervising Officer' shall be deemed
by a person who is not entitled to the use of the name 'Architect'   to have been deleted.
under and in accordance with the Architects (Registration) Acts
1931 to 1969. If so, the expression 'Architect' shall be deemed to
have been deleted throughout this Tender including the

Tender NSC/1   **Schedule 1**

7. Execution of Main Contract:

*is/is to be

*under hand/under seal

8. Main Contract Conditions –
alternative etc. provisions:

Architect/Supervising Officer: *Article 3A/Article 3B

WITHOUT Quantities editions only: *Article 4A/Article 4B

master programme: Clause 5·3·1·2 *deleted/not deleted

Works insurance: *Clause 22A   22B   22C

insurance: Clause 21·2·1 Provisional sum *included/not included

9. Main Contract Conditions – any changes from printed Standard Form identified in item 5:

ARCHITECT - STAGE 2

10. Main Contract: Appendix and entries therein

	Clause etc	
Statutory tax deduction scheme – Finance (No.2) Act 1975	Fourth recital and 31	Employer at Date of Tender *is a 'contractor'/is not a 'contractor' for the purposes of the Act and the Regulations.
Settlement of disputes – Arbitration	Article 5·1	Articles 5·1·4 and 5·1·5 apply (see Article 5·1·6)
Date for Completion	1·3 [j]	
Defects Liability Period (if none other stated is 6 months from the day named in the Certificate of Practical Completion of the Works).	17·2	
Insurance cover for any one occurrence or series of occurrences arising out of one event	21·1·1	£
Percentage to cover professional fees	22A	
Date of Possession	23·1	
Liquidated and Ascertained Damages	24·2	at the rate of £_____ per _____

130

10. Main Contract: Appendix *continued*

	Clause	
Period of delay:	28·1·3	
(i) by reason of loss or damage caused by any one of the Clause 22 Perils	28·1·3·2	
(ii) for any other reason	28·1·3·1, 28·1·3·3 to ·3·7	
Period of Interim Certificates (if none stated is one month)	30·1·3	
Retention Percentage (if less than five per cent)	30·4·1·1	
Period of Final Measurement and Valuation (if none stated is 6 months from the day named in the Certificate of Practical Completion of the Works)	30·6·1·2	
Period for issue of Final Certificate (if none stated is 3 months)	30·8	
Work reserved for Nominated Sub-Contractors for which the Contractor desires to tender	35·2	
†Fluctuations	37	*Clause 38/Clause 39/Clause 40
Percentage addition	*38·7/39·8	
†Formula Rules	40·1·1·1	
	rule 3	Base Month _____ 19___
	rule 3	Non-Adjustable Element |k|_____ % (not to exceed 10%)
	rule 10	Part I/Part II of Section 2 of the Formula Rules is to apply.

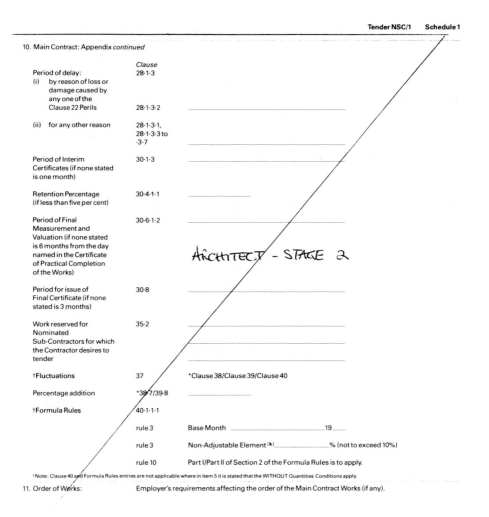

ARCHITECT - STAGE 2

†Note: Clause 40 and Formula Rules entries are not applicable where in item 5 it is stated that the WITHOUT Quantities Conditions apply.

11. Order of Works: Employer's requirements affecting the order of the Main Contract Works (if any).

Tender NSC/1   **Schedule 1**

12. Location and type of access:

13. Obligations or restrictions imposed by the Employer not covered by Main Contract Conditions:
    (e.g. in Preliminaries in the Contract Bills)[1]

ARCHITECT - STAGE 2

14. Other relevant information: (if any)

Signed by or on behalf   NOMINATED SUB-CONTRACTOR - STAGE 3
of the Sub-Contractor   Information noted _____

Page 6

132

Fluctuations
See Tender NSC/1 page 3, item 2

Notes:
Complete column (1) where Sub-Contract NSC/4, clause 35 applies.
Complete columns (1)-(4) where Sub-Contract NSC/4, clause 36 etc.
applies.

The 'Date of Tender'[n] for the purposes of Sub-Contract NSC/4,
clauses 35/36.

is _ARCHITECT - STAGE 2_

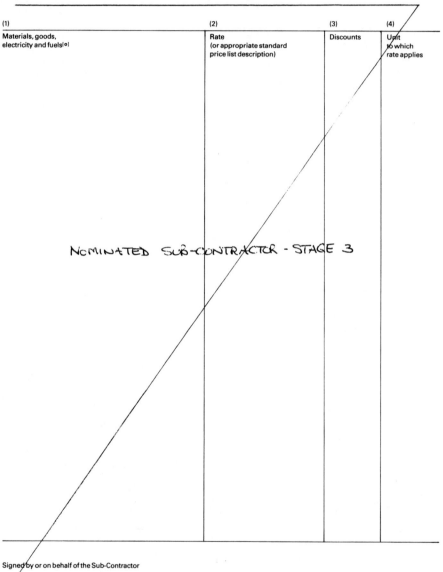

(1) Materials, goods, electricity and fuels[o]	(2) Rate (or appropriate standard price list description)	(3) Discounts	(4) Unit to which rate applies

NOMINATED SUB-CONTRACTOR - STAGE 3

Signed by or on behalf of the Sub-Contractor

Date

Page 7

Tender NSC/1   Schedule 1   Appendix (B)

Fluctuations
See Tender NSC/1 page 3, item 2

Note: Architect and Sub-Contractor to complete this Appendix as appropriate where Sub-Contract NSC/4, clause 37 applies.

ARCHITECT — STAGE 2

1. 37·1 – Nominated Sub-Contract
   Formula Rules are those dated _____ 198__
   *Part I/Part III of these Rules applies
2. 37·3·3 and 3·4 – Non-Adjustable Element[p] _____ % (not to exceed 10%)

3. 37·4 – List of Market Prices   *NOMINATED SUB-CONTRACTOR — STAGE 3*

4. Nominated Sub-Contract Formula Rules
   rule 3 (Definition of Balance of Adjustable Work)
   Any measured work not allocated to a Work Category

NOMINATED SUB-CONTRACTOR — STAGE 3

Base Month (rule 3) _____ *ARCHITECT — STAGE 2*

Date of Tender (rule 3) _____

rule 8 Method of dealing with 'Fix-only' work
NOMINATED SUB-CONTRACTOR — STAGE 3
rule 11(a) Part I only: the Work Categories applicable to the Sub-Contract Works

rule 43 Part III only: Weightings of labour and materials – Electrical Installations or Heating, Ventilating and Air Conditioning Installations[q]

ARCHITECT — STAGE 2

	Labour	Materials
Electrical	_____ %	_____ %
Heating, Ventilating and Air Conditioning[r]	_____ %	_____ %
	_____ %	_____ %

rule 61a Adjustment shall be effected
*upon completion of manufacture of all fabricated components
*upon delivery to site of all fabricated components

rule 64 Part III only: Structural Steelwork Installations:
(i) Average price per tonne of steel delivered to fabricator's work
£ _____
NOMINATED SUB-CONTRACTOR — STAGE 3
(ii) Average price per tonne for erection of steelwork
£ _____

rule 70a Catering Equipment Installations:
apportionment of the value of each item between
(i) Materials and shop fabrication   £ _____

(ii) supply of factor items   £ _____

(iii) site installations   £ _____

Signed by or on behalf of the Sub-Contractor

Date

     *Delete as applicable

134

Schedule 2: Particular Conditions

Note: When the Contractor receives Tender NSC/1 together with the Architect/Supervising Officer's preliminary notice of nomination under clause 35·7·1 of the Main Contract Conditions then the Contractor has to settle and complete any of the particular conditions which remain to be completed in this Schedule in

agreement with the proposed Sub-Contractor. The completed Schedule should take account not only of the preliminary indications of the Sub-Contractor stated therein, but also of any particular conditions or requirements of the Contractor which he may wish to raise with the Sub-Contractor.

1.A   Any stipulation as to the period/periods when Sub-Contract Works can be carried out on site:[s]

to be between   ARCHITECT - STAGE 2   and

Period required by Architect to approve drawings after submission

1.B   Preliminary programme details[t] (having regard to the information provided in the invitation to tender)

Periods required:

(1)   for submission of all further sub-contractors drawings etc. *(co-ordination, installation, shop or builders' work, or other as appropriate)*[u]

NOMINATED SUB-CONTRACTOR - STAGE 3

(2)   for execution of Sub-Contract Works: off-site

on-site

Notice required to commence work on site

1.C   Agreed programme details (including sub-contract completion date: see also Sub Contract NSC/4, clause 11·1)[v]

BY AGREEMENT
CONTRACTOR AND NOMINATED
SUB-CONTRACTOR - STAGE 7

2.   Order of Works to follow the requirements, if any, stated in Schedule 1, item 11[w]

3.A Attendance proposals (other than †general attendance).[×]

(a) Special scaffolding or scaffolding additional to the Contractor's standing scaffolding.

(b) The provision of temporary access roads and hardstandings in connection with structural steelwork, precast concrete components, piling, heavy items of plant and the like.

NOMINATED SUB-CONTRACTOR - STAGE 3

(c) Unloading, distributing, hoisting and placing in position giving in the case of significant items the weight and/or size. (To be at the risk of the Sub-Contractor).

(d) The provision of covered storage and accommodation including lighting and power thereto.

(e) Power supplies giving the maximum load.

(f) Maintenance of specific temperature or humidity levels.

(g) Any other attendance not included under (a) to (f) or as †general attendance under Sub-Contract NSC/4, paragraph 27·1·1.

†Note: For general attendance see clause 27·1·1 of Sub-Contract NSC/4 which states: "General attendance shall be provided by the Contractor free of charge to the Sub-Contractor and shall be deemed to include only use of the Contractor's temporary roads, pavings and paths, standing scaffolding, standing power operated hoisting plant, the provision of temporary lighting and water supplies, clearing away rubbish, provision of space for the Sub-Contractor's own offices and for the storage of his plant and materials and the use of messrooms, sanitary accommodation and welfare materials." See SMM, 6 edn., B.9.2.

136

3.B   (y)

CONTRACTOR AND
NOMINATED SUB-CONTRACTOR
BY AGREEMENT
— STAGE 7

4.   Insurance[z]

5.A   Employment of Labour – Special Conditions or Agreements[aa]

NOMINATED SUB-CONTRACTOR
— STAGE 3

5.B   Employment of Labour – Special Conditions or Agreements[bb]

CONTRACTOR AND NOMINATED SUB-CONTRACTOR
BY AGREEMENT — STAGE 7

6.   The Adjudicator is:[cc]

      The Trustee – Stakeholder is:[cc]   *CONTRACTOR - STAGE 7*

7.   Finance (No. 2 Act 1975 – Statutory Tax Deduction Scheme[dd]

      1. The Contractor* is/is not entitled to be paid by the Employer without the statutory deduction referred to in the above Act or such other deduction as may be in force;

      2. The Sub-Contractor* is/is not entitled to be paid by the Contractor without the above-mentioned statutory deduction or such other deduction;

      3. The evidence to be produced to the Contractor for the verification of the Sub-Contractor's tax certificate (expiry date   19   ) will be:

                    CONTRACTOR AND NOMINATED
                    SUB-CONTRACTOR BY AGREEMENT
                    - STAGE 7

8.   Value Added Tax – Sub-Contract NSC/4
      Clause 19A/19B[ee] (alternative VAT provisions) will apply.

9.A   Any other matters (including any limitation on working hours).[ff]

            NOMINATED SUB-CONTRACTOR - STAGE 3

9.B   [gg]

            CONTRACTOR AND NOMINATED
            SUB-CONTRACTOR
            BY AGREEMENT - STAGE 7

10.   Any matters agreed by the Architect/Supervising Officer and Sub-Contractor before preliminary notice of nomination.[s]

            ARCHITECT - STAGE 6

11.   Sub-Contract NSC/4 Edition as identified in Schedule 1, Item 1 to be executed under hand/under seal[ee] forthwith after Architect/Supervising Officer's nomination on Nomination NSC/3.
      CONTRACTOR AND NOMINATED SUB-CONTRACTOR BY AGREEMENT - STAGE 7

The above Particular Conditions are agreed

Signed by or on behalf of   Date
the Sub-Contractor   *NOMINATED SUB-CONTRACTOR*
                     - STAGE 7

Signed by or on behalf   *CONTRACTOR - STAGE 7*   Date
of the Main Contractor

Explanatory notes

The first page of the Tender NSC/1 has the description of the Main Contract Works, Location and Sub-contract Works. The Architect will take care here to have these descriptions very specific, more especially with the Sub-contract Works description. It is possible that a design requirement is necessary to suit a performance specification, such as, as a simple example, curtain walling able to withstand wind pressures and expansion and contraction parameters. The Nominated Sub-contractor has to provide a Warranty in Agreement NSC/2 for his design, if design is required, and for the satisfaction of any performance specification, if it is required. This Warranty is restricted to the description of the works included in the Tender, thus that description must be explicit on all the points required. It is usual, therefore, that lack of space on the form will require a separate sheet to be attached to the Tender and reference made to it in the description of the sub-contract works.

The Tender may be prepared by the Nominated Sub-contractor in accordance with his own, or issued, drawings, specifications, Bills of Quantities or Schedule of Rates. The non-applicable alternatives are deleted, every document and drawing is numbered for identification and entered in the form. Note that these documents, drawings, etc., and Tender, become the Sub-contract Documents referred to in NSC/4.

The Tender amount is stated as either a Sub-contract Sum or a Tender Sum and the alternative will be deleted. A Sub-contract Sum is a lump sum offer and thus adjusted by variations. The Tender Sum is not lump sum and is subject to re-measurement and valuation to achieve the final account or Ascertained Final Contract Sum.

The Architect may wish to use Tender NSC/1 for competitive offers and could therefore decide at Stage 1 which form of offer he requires and so delete one of the alternatives at that stage.

Note that NSC/4, the Sub-contract Conditions, states that Bills of Quantities become a contract document and have to be prepared in accordance with SMM6. Therefore, the Nominated Sub-contractor who prepares his own Bill of Quantities has to be careful in their preparation, and equally so must the Architect be in accepting them.

The offer and the daywork percentages have to include a 2½ per cent discount to the Main Contractor which, of course, is applicable only if he makes payments within the prescribed period to the Nominated Sub-contractor.

Conditions 1 to 5 on page 2 are self-explanatory, but note in Condition 4 that a time has to be inserted by the Nominated Sub-contractor.

Schedule 1

Since the Architect completes this Schedule it is his sole responsibility to be very sure that all the facts are transferred here correctly. This schedule governs the Sub-contract NSC/4 and therefore any errors here will stand even if different from the Main Contract. If tenders for Nominated Sub-contract Works are being invited either before, or at the same time as, the Main Contract, there is the additional problem of arriving at the exact dates and terms required to suit both the Main Contract and the Nominated Sub-contracts.

Item 1 This states the Sub-contract Conditions to be NSC/4, which are not to be amended, and records the Main Contract Conditions.

Item 2 One of the three fluctuation clauses must be chosen and the other two deleted. If none has been identified, then Clause 35 will operate.

Item 3 Any items in the Main Contract Appendix on Pages 4 and 5 Item 10 which are not to apply to the Sub-contract, will be entered here.

Item 4 This is a full description of the Main Contract Works.

Item 5 This details the version of the Standard Conditions and any revision thereof. If Sectional Completion is to operate, then the item should be so amended.

Item 6 It is most helpful if all the documents are lodged at the same address. Note deletion required.

Item 7 The form of execution is to be stated here. Note also the deletion indicating whether or not the Main Contract has already been executed.

Item 8 Here deletion should have been made to accord with the options chosen in the Main Contract.

Item 9 Any amendment or supplementary clause from or to the Standard Form, will be detailed here.

Item 10 Details of the Main Contract Appendix are transferred here and the Contractor should check their validity when he receives Tender NSC/1 and the Preliminary Notice of Nomination. Remember that the Tender Schedule 1 governs the Sub-contract NSC/4, as clearly stated in Clause 2.2 of Sub-contract NSC/4.

Item 11 This must be an exact duplicate of the information given to the Main Contractor in the Preliminaries Bill, as far as it is applicable. A copy extract from the Preliminaries Bill might well be suitable.

Item 12 Again this refers to the Preliminaries Bill.

Item 13 The sidenote suggests here that an unpriced copy of the Preliminaries Bill could be attached to Tender NSC/1, and if this is done it will be so noted against this item. Remember that the Sub-contractor can reasonably object to a variation of the entries made here.

Item 14 Again the Preliminaries Bill being attached would cover this item, if not, they would have to be noted in detail.

Schedule 1: Appendix A

This Appendix is completed by the Sub-contractor but only when Clauses 35 or 36 apply, otherwise the page would be struck out. Note on the Appendix the instructions as to the proper columns to be used. The list in Column 1 is completed and fluctuations will apply only to the materials and goods and electricity entered by the Sub-contractor.

The 'date of tender' will be inserted by the Architect and he will exercise care in establishing this date, as being the date ten days before the date fixed for lodgement of tenders. If, however, the completion of Tender NSC/1 follows a preliminary tender already obtained, the date would have to be set to accord with when the Contract Sum or Tender Sum was prepared, and so reflect the base date for fluctuations.

'Fuels' is an option for the Employer which will be struck out or left in by the Architect, at the same time as he enters the date of tender. This option is stated in Clause 35.2.1 or 36.3.1 of Sub-contract NSC/4.

Schedule 1: Appendix B

This Appendix will be used only where Clause 37 (Formula adjustment) has been selected, and if Appendix A operates then this page will have a line drawn through it to indicate that it does not apply.

Item 1   The Edition of the Formula Rules is identified by date and either Part I or Part III is selected. Part I is the Work Category Method and Part III applies to specialist engineering installations.

Item 2   This applies only to the Local Authorities with Quantities Edition where the percentage operates to curtail the formula adjustment. The Private Edition of the Standard Form has the formula adjustment paid in full.

Item 3   The list inserted here by the Sub-contractor is of articles manufactured outside the United Kingdom, imported, and without any further processing fixed directly into the works, and which he wishes to exclude from formula adjustment but have subject to direct variation in market price. (*See* Clause 37A of Sub-contract NSC/4).

Item 4   The Sub-contractor will fill in here any items which he considers not able to be allocated to a work category. There are 48 categories listed in Appendix A to the rules.

The 'base month' to be inserted by the Architect would normally be the calendar month before that in which the tender is due to be returned, unless any other arrangement is stated. It must be pointed out however, that if a Preliminary tender had been obtained prior to NSC/1 being formalized, then the base month would be that preceding the month in which the preliminary tender was prepared. Both the Architect and the sub-contractor would each in turn check that the entry was correct.

The 'date of tender' to be inserted by the Architect, would normally be a date ten days before the date for return of tenders. Once again, in the case of a preliminary tender, the date of tender would have to reflect when the Contract Sum or Tender Sum was actually prepared.

There are three options open to a tenderer as to how fix-only work will be treated in formula adjustment, namely

(a) Inclusion in an appropriate work category,
(b) Inclusion in the Balance of Adjustable Work,
(c) Application of the 'fix-only' index as Appendix B Sub-contract Formula Rules.

One of these options must be entered.

Rule 11(a), when Part I only applies, there has to be entered here that a schedule is attached to the Tender which shows each item of the Sub-contractor's bills or schedules, classified as to work category, fix-only work, specialist engineering work, provisional sum work, balance of adjustable work, or work excluded from formula adjustment.

Rule 43 is entered by the Architect and Rules 61a, 64 and 70a by the Sub-contractor, all of which are self-explanatory.

Schedule 2: Particular Conditions

The form indicates which individuals should complete the relevant parts. Note that Items 1C, 3B, 4, 5B and 9B are settled and completed, by both the Main Contractor and the Nominated Sub-contractor, thus superseding the preliminary information given in Items 1A, 1B, 5A and 9A, which entries are then deleted. This then leaves the remainder, after the deletions, as firm final statements which will eventually form part of the Sub-contract.

Item 1A   If the preparation of Tender NSC/1 comes after the Main Contract is let, then the necessary information can be obtained from the Main Contractor or his programme. When the Tender is being requested at the same time or in advance of the Main Contract, then the Architect can only use his judgement, as to the possible contract period for the Main Contract.

If design forms part of the Sub-contract, the Architect will state here the time he requires to examine the drawings for approval.

Item 1B   The periods filled in here by the Sub-contractor are, of course, governed by whether or not there is a design element in the Tender, and this should be made clear. Each type of drawing and its purpose will be entered, together with the time required. As this is the stage at which the Sub-contractor is giving his preliminary information, it would be more effective if the time periods were stated as a start and finish date, to enable the Contractor and Sub-contractor to complete 1C in detail.

Item 1C   The programme details to be settled will include the information given in 1A and 1B, which will be notice to start work on site, start date and completion date, and the periods required for the various categories of drawings.

These dates or times constitute the base from which variations, extensions of time and direct loss and/or expense will operate, when Sub-contract NSC/4 is executed.

Again, it should be reiterated that 1A and 1B will be deleted and the deletions initialled by both parties.

Item 2   If any additional requirements are necessary because of the Sub-contract works, they would be added here. It is possible that any special order of work required might be entered here by the Architect at Stage 2.

Item 3A   This item comprises a detailed list of the special attendance required by the Sub-contractor, and it is additional to the free general attendance listed at the foot of the page.

This list is again preliminary information from the Sub-contractor, which never-the-less should be completely detailed. The Contractor has to price for these items accurately when he fills in the Main Contract Bills.

1 – As the clause suggests, the Sub-contractor has the free use of the Contractor's standing scaffolding for as long as the scaffold will be in place to suit the Contractor. Therefore, special scaffolding could be:

(a) Adaptation, or extension of time of the standing scaffolding, or
(b) Special separate scaffolding.

The details of the scaffolding are important; where is it to be, how high, for how long in position, is it to be adapted from time to time, is it mobile, or what type construction?

2 – The Sub-contractor will have the free use of the Contractor's temporary roads and hardstandings for as long as the Contractor uses them. Therefore, in theory, the Sub-contractor may require an extension of the roads, using them for a longer time, increasing their load capacity, or special roads of his own. In practical terms, at this stage the Sub-contractor cannot know what the Contractor will be providing, so his requirements must be stated both in total and in detail.

Ideally, the Sub-contractor could state the expected type of vehicles delivering his goods and plant to site, giving details of the overall size and total loaded weight. He could also state the weight and size of plant expected to use a hardstanding. Consider the structural steelwork contractor with long heavy loads and possibly a crane on site.

3 – Once again the Sub-contractor has the free use of power-operated hoists or lifts supplied by the Contractor for his own use, and only for so long as the Contractor requires them.

If the Contractor has to unload materials prior to the Sub-contractor's labour force being on site, then details of the timing, nature and packaging of the materials must be given.

If a further stage is required, and goods or materials are to be distributed, especially throughout several floors, details of the amounts, sizes, weights and type of packaging, should be given, if possible.

Hoisting will be carried out by the Contractor if he has adequate facilities on site; thus, details of the types, sizes, weights and amounts of goods to be hoisted should be given. If the proposed hoisting facilities would prove inadequate, these details will enable the Contractor to judge what will be necessary.

Placing in position is normally only for very heavy items and their descriptive details will have been included previously.

4 – The Sub-contractor gets free space provided only for his offices, plant and material storage. Should the Sub-contractor not wish to provide his own offices and covered storage, details of what would be required, the period required for, construction, etc., would need to be given for the Main Contractor to supply them. Also the requirements as to lighting and power need to be known.

5 – Power is not supplied as a general attendance item and the Sub-contractor has to state where on site he requires it and of what

phase or voltage. If several outlets are required, they must be indicated as to position, voltage and duration of use.

6 – It is possible that some materials might require special levels of temperature and or humidity during storage, during installation, after installation, or any combination of these alternatives. The levels, or minimum to maximum levels, must be given and the times during which these levels have to be maintained.

7 – Any other attendances outwith general attendance, not already mentioned, but which are to suit some specialist trade, should be included here.

Item 3B   When details of the special attendance have been agreed between the Sub-contractor and the Contractor, this entry is intended to record any alteration to the attendances required by the Sub-contractor in 3A.

Item 4   The Contractor and the Sub-contractor agree the insurance particulars, including limits to indemnity to be covered by the Sub-contractor, and all to be in accordance with Clause 7 of Sub-contract NSC/4.

Item 5A   Details to be given here of any special conditions or agreements affecting the employment of labour. The item is deleted and initialled by both parties after completion of 5B.

Item 5B   The agreement between both parties, in 5A, is entered here by the Contractor and thereafter 5A deleted and initialled.

Item 6   After agreement with the Sub-contractor, the Contractor to fill in the names and addresses of the person or firm they jointly have agreed, will ajudicate or arbitrate on any differences. Also that of a person, firm or bank who will hold monies in trust, to be disposed of according to Clause 24.5 of Sub-contract NSC/4.

Item 7   The three items here are agreed between the Sub-contractor and the Contractor at Stage 7 and filled in by the Contractor. Without these declarations Clauses 20A and 20B of Sub-contract NSC/4 cannot operate properly.

Item 8   The VAT alternative is agreed between the two parties and entered by the Contractor. 19A will operate where the sub-contractor prepares authenticated receipts, or where the Contractor prepares tax documents in lieu of authenticated receipts and the Sub-contractor *does not* consent to the use of this method. 19B is an alternative where the Contractor is allowed to prepare tax documents in lieu of authenticated receipts and the Sub-contractor consents. Whichever alternative is used, the other will be deleted.

Item 9A   The Sub-contractor can propose here any other matters which he feels an agreement with the Contractor would be advantageous or necessary, before he is nominated. These may be matters which affect a specialist trade or limitations to working hours. The possibility of annual or public holiday entitlements not coinciding with those of the Contractor and he wishing to work while the Contractor is absent from site. The Sub-contractor may have had problems on other contracts which some agreement here could obviate.

Item 9B   When the matters in 9A are settled and agreed with the Contractor, the details are entered here by the Contractor and 9A deleted and initialled by both parties.

Item 10   The Architect fills in any agreed matters between himself and the Sub-contractor at Stage 6 prior to his issue of a preliminary notice of nomination.

Item 11   Agreement is made here as to whether or not Sub-contract NSC/4 is to be executed either under hand or under seal.

Agreement NSC/2

JCT

JCT Standard Form of Employer/Nominated Sub-Contractor Agreement

Agreement between a Sub-Contractor to be nominated for Sub-Contract Work in accordance with clauses 35·6 to 35·10 of the Standard Form of Building Contract for a main contract and the Employer referred to in the main contract.

Main Contract Works:

Location: *ARCHITECT - STAGE 2*

Sub-Contract Works:

This Agreement

The date to be inserted here must be the date when the Tender NSC/1 is signed as 'approved' by the Architect/Supervising Officer on behalf of the Employer

is made the_____day of_____19 _____

between _____

of or whose registered office is situated at _____

_____

(hereinafter called 'the Employer') and

_____

of or whose registered office is situated at _____

_____

(hereinafter called 'the Sub-Contractor')

Whereas

First   the Sub-Contractor has submitted a tender on Tender NSC/1 (hereinafter called 'the Tender') on the terms and conditions in that Tender to carry out Works (referred to above and hereinafter called 'the Sub-Contract Works') as part of the Main Contract Works referred to above to be or being carried out on the terms and conditions relating thereto referred to in Schedule 1 of the Tender (hereinafter called 'the Main Contract');

Second   the Employer has appointed

_____

to be the Architect/Supervising Officer for the purposes of the Main Contract and this Agreement (hereinafter called 'the Architect/Supervising Officer' which expression as used in this Agreement shall include his successors validly appointed under the Main Contract or otherwise before the Main Contract is operative).

Third   the Architect/Supervising Officer on behalf of the Employer has approved the Tender and intends that after agreement between the Contractor and Sub-Contractor on the Particular Conditions in Schedule 2 thereof an instruction on Nomination NSC/3 shall be issued to the Contractor for the Main Contract (hereinafter called 'the Main Contractor') nominating the Sub-Contractor to carry out and complete the Sub-Contract Works on the terms and conditions of the Tender;

Fourth   nothing contained in this Agreement nor anything contained in the Tender is intended to render the Architect/Supervising Officer in any way liable to the Sub-Contractor in relation to matters in the said Agreement and Tender.

© 1980 RIBA Publications Ltd

Now it is hereby agreed

1·1   The Sub-Contractor shall, after the Architect/Supervising Officer has issued his preliminary notice of nomination under clause 35·7·1 of the Main Contract Conditions, forthwith seek to settle with the Main Contractor the Particular Conditions in Schedule 2 of the Tender.

1·2   The Sub-Contractor shall, upon reaching agreement with the Main Contractor on the Particular Conditions in Schedule 2 of the Tender and after that Schedule is signed by or on behalf of the Sub-Contractor and the Main Contractor, immediately through the Main Contractor so inform the Architect/Supervising Officer.

2·1   The Sub-Contractor warrants that he has exercised and will exercise all reasonable skill and care in

  ·1   the design of the Sub-Contract Works insofar as the Sub-Contract Works have been or will be designed by the Sub-Contractor; and

  ·2   the selection of materials and goods for the Sub-Contract Works insofar as such materials and goods have been or will be selected by the Sub-Contractor; and

  ·3   the satisfaction of any performance specification or requirement insofar as such performance specification or requirement is included or referred to in the description of the Sub-Contract Works included in or annexed to the Tender.

Nothing in clause 2·1 shall be construed so as to affect the obligations of the Sub-Contractor under Sub-Contract NSC/4 in regard to the supply under the Sub-Contract of workmanship, materials and goods.

2·2   ·1   If, after the date of this Agreement and before the issue by the Architect/Supervising Officer of the instruction on Nomination NSC/3 under clause 35·10·2 of the Main Contract Conditions, the Architect/Supervising Officer instructs in writing that the Sub-Contractor should proceed with
  ·1   the designing of, or
  ·2   the proper ordering or fabrication of any materials or goods for
the Sub-Contract Works the Sub-Contractor shall forthwith comply with the instruction and the Employer shall make payment for such compliance in accordance with clauses 2·2·2 to 2·2·4.

  ·2   No payment referred to in clauses 2·2·3 and 2·2·4 shall be made after the issue of Nomination NSC/3 under clause 35·10·2 of the Main Contract Conditions except in respect of any design work properly carried out and/or materials or goods properly ordered or fabricated in compliance with an instruction under clause 2·2·1 but which are not used for the Sub-Contract Works by reason of some written decision against such use given by the Architect/Supervising Officer before the issue of Nomination NSC/3.

  ·3   The Employer shall pay the Sub-Contractor the amount of any expense reasonably and properly incurred by the Sub-Contractor in carrying out work in the designing of the Sub-Contract Works and upon such payment the Employer may use that work for the purposes of the Sub-Contract Works but not further or otherwise.

  ·4   The Employer shall pay the Sub-Contractor for any materials or goods properly ordered by the Sub-Contractor for the Sub-Contract Works and upon such payment any materials and goods so paid for shall become the property of the Employer.

  ·5   If any payment has been made by the Employer under clauses 2·2·3 and 2·2·4 and the Sub-Contractor is subsequently nominated in Nomination NSC/3 issued under clause 35·10·2 of the Main Contract Conditions to execute the Sub-Contract Works the Sub-Contractor shall allow to the Employer and the Main Contractor full credit for such payment in the discharge of the amount due in respect of the Sub-Contract Works.

Delay in supply of
information and in
performance by
Sub-Contractor

3·1   The Sub-Contractor will not be liable under clauses 3·2, 3·3 or 3·4 until the Architect/Supervising Officer has issued his instruction on Nomination NSC/3 under clause 35·10·2 of the Main Contract Conditions nor in respect of any revised period of time for delay in carrying out or completing the Sub-Contract Works which the Sub-Contractor has been granted under clause 11·2 of Sub-Contract NSC/4.

3.2   The Sub-Contractor shall so supply the Architect/Supervising Officer with information (including drawings) in accordance with the agreed programme details or at such time as the Architect/Supervising Officer may reasonably require so that the Architect/Supervising Officer will not be delayed in issuing necessary instructions or drawings under the Main Contract, for which delay the Main Contractor may have a valid claim to an extension of time for completion of the Main Contract Works by reason of the Relevant Event in clause 25·4·6 or a valid claim for direct loss and/or expense under clause 26·2·1 of the Main Contract Conditions.

3·3   The Sub-Contractor shall so perform his obligations under the Sub-Contract that the Architect/ Supervising Officer will not by reason of any default by the Sub-Contractor be under a duty to issue an instruction to determine the employment of the Sub-Contractor under clause 35·24 of the Main Contract Conditions provided that any suspension by the Sub-Contractor of further execution of the Sub-Contract Works under clause 21·8 of Sub-Contract NSC/4 shall not be regarded as a 'default by the Sub-Contractor' as referred to in clause 3·3.

3·4   The Sub-Contractor shall so perform the Sub-Contract that the Contractor will not become entitled to an extension of time for completion of the Main Contract Works by reason of the Relevant Event in clause 25·4·7 of the Main Contract Conditions.

Architect's direction on value of Sub-Contract Work in Interim Certificates – information to Sub-Contractor

4   The Architect/Supervising Officer shall operate the provisions of *clause 35·13·1 of the Main Contract Conditions.

Employer's duty – final payment for Sub-Contract Works

5·1   The Architect/Supervising Officer shall operate the provisions in †clauses 35·17 to 35·19 of the Main Contract Conditions.

Discharge of final payment to Sub-Contractor – Sub-Contractor's obligations

5·2   After due discharge by the Contractor of a final payment under clause 35·17 of the Main Contract Conditions the Sub-Contractor shall rectify at his own cost (or if he fails so to rectify, shall be liable to the Employer for the costs referred to in clause 35·18 of the Main Contract Conditions) any omission, fault or defect in the Sub-Contract Works which the Sub-Contractor is bound to rectify under Sub-Contract NSC/4 after written notification thereof by the Architect/Supervising Officer at any time before the issue of the Final Certificate under clause 30·8 of the Main Contract Conditions.

5·3   After the issue of the Final Certificate under the Main Contract Conditions the Sub-Contractor shall in addition to such other responsibilities, if any, as he has under this Agreement, have the like responsibility to the Main Contractor and to the Employer for the Sub-Contract Works as the Main Contractor has to the Employer under the terms of the Main Contract relating to the obligations of the Contractor after the issue of the Final Certificate.

Architect's instructions – duty to make a further nomination – liability of Sub-Contractor

6   Where the Architect/Supervising Officer has been under a duty under clause 35·24 of the Main Contract Conditions except as a result of the operation of clause 35·24·6 to issue an instruction to the Main Contractor making a further nomination in respect of the Sub-Contract Works, the Sub-Contractor shall indemnify the Employer against any direct loss and/or expense resulting from the exercise by the Architect/Supervising Officer of that duty.

Architect's certificate of non-discharge by Contractor – payment to Sub-Contractor by Employer

7·1   The Architect/Supervising Officer and the Employer shall operate the provisions in regard to the payment of the Sub-Contractor in clause 35·13 of the Main Contract Conditions.

7·2   If, after paying any amount to the Sub-Contractor under clause 35·13·5·3 of the Main Contract Conditions, the Employer produces reasonable proof that there was in existence at the time of such payment a petition or resolution to which clause 35·13·5·4·4 of the Main Contract Conditions refers, the Sub-Contractor shall repay on demand such amount.

Conditions in contracts for purchase of goods and materials – Sub-Contractor's duty

8   Where ‡clause 2·3 of Sub-Contract NSC/4 applies, the Sub-Contractor shall forthwith supply to the Contractor details of any restriction, limitation or exclusion to which that clause refers as soon as such details are known to the Sub-Contractor.

Conflict between Tender and Agreement

9   If any conflict appears between the terms of the Tender and this Agreement, the terms of this Agreement shall prevail.

Arbitration

10·1   In case any dispute or difference shall arise between the Employer or the Architect/Supervising Officer on his behalf and the Sub-Contractor, either during the progress or after the completion or abandonment of the Sub-Contract Works, as to the construction of this Agreement, or as to any matter or thing of whatsoever nature arising out of this Agreement or in connection therewith, then such dispute or difference shall be and is hereby referred to the arbitration and final decision of a person to be agreed between the parties, or, failing agreement within 14 days after either party has given to the other a written request to concur in the appointment of an Arbitrator, a person to be appointed on the request of either party by the President or a Vice President for the time being of the Royal Institute of British Architects.

---

*Note: Clause 35·13·1 requires that the Architect/Supervising Officer, upon directing the Main Contractor as to the amount included in any Interim Certificates in respect of the value of the Nominated Sub-Contract Works issued under clause 30 of the Main Contract Conditions, shall forthwith inform the Sub-Contractor in writing of that amount.

†Note: Clause 35·17 deals with final payment by the Employer for the Sub-Contract Works prior to the issue of the Final Certificate under the Main Contract Conditions.

‡Note: Clause 2·3 deals with specified supplies and restrictions etc. in the contracts of sale for such supplies.

Agreement NSC/2

10·2 ·1 Provided that if the dispute or difference to be referred to arbitration under this Agreement raises issues which are substantially the same as or connected with issues raised in a related dispute between the Employer and the Contractor under the Main Contract or between the Sub-Contractor and the Contractor under Sub-Contract NSC/4 or NSC/4a or between the Employer and any other nominated Sub-Contractor under Agreement NSC/2 or NSC/2a or between the Employer and any Nominated Supplier whose contract of sale with the Main Contractor provides for the matters referred to in clause 36·4·8 of the Main Contract Conditions, and if the related dispute has already been referred for determination to an Arbitrator, the Employer and the Sub-Contractor hereby agree that the dispute or difference under this Agreement shall be referred to the Arbitrator appointed to determine the related dispute; and such Arbitrator shall have power to make such directions and all necessary awards in the same way as if the procedure of the High Court as to joining one or more defendants or joining co-defendants or third parties was available to the parties and to him.

10·2 ·2 Save that the Employer or the Sub-Contractor may require the dispute or difference under this Agreement to be referred to a different Arbitrator (to be appointed under this Agreement) if either of them reasonably considers that the Arbitrator appointed to determine the related dispute is not appropriately qualified to determine the dispute or difference under this Agreement.

10·2 ·3 Clauses 10·2·1 and 10·2·2 shall apply unless in the Appendix to the Main Contract Conditions the words "Articles 5·1·4 and 5·1·5 apply" have been deleted.

10·3 Such reference shall not be opened until after Practical Completion or alleged Practical Completion of the Main Contract Works or termination or alleged termination of the Contractor's employment under the Main Contract or abandonment of the Main Contract Works, unless with the written consent of the Employer or the Architect/Supervising Officer on his behalf and the Sub-Contractor.

10·4 The award of such Arbitrator shall be final and binding on the parties.

10·5* Whatever the nationality, residence or domicile of the Employer, the Contractor, any Sub-Contractor or supplier or the Arbitrator, and wherever the Works or any part thereof are situated, the law of England shall be the proper law of this Agreement and in particular (but not so as to derogate from the generality of the foregoing) the provisions of the Arbitration Acts 1950 (notwithstanding anything in S·34 thereof) to 1979 shall apply to any arbitration under Clause 10 wherever the same, or any part of it, shall be conducted.

*Where the parties do not wish the proper law of the contract to be the law of England and/or do not wish the provisions of the Arbitration Acts 1950 to 1979 to apply to any arbitration under the   Contract held under the procedural law of Scotland (or other country) appropriate amendments to Clause 10·5 should be made.

Signed by or on behalf of the Sub-Contractor [g1] _NOMINATED SUB-CONTRACTOR – STAGE 3_

in the presence of:

Signed by or on behalf of the Employer [g1] _EMPLOYER – STAGE 5_

in the presence of:

Signed, sealed and delivered by [g2]/The common seal of [g3]: _____

in the presence of [g2]/was hereunto affixed in the presence of [g3]:

Signed, sealed and delivered by [g2]/The common seal of [g3]: _____

in the presence of [g2]/was hereunto affixed in the presence of [g3]:

Footnotes

[g1] For use if Agreement is executed under hand.

[g2] For use if executed under seal by an individual or firm or unincorporated body.

[g3] For use if executed under seal by a company or other body corporate.

Page 4

Explanatory notes

The Agreement NSC/2 is properly executed at the same time as and in conjunction with Tender NSC/1. Again, the form has its parts indicated to show where the various parties enter the information of affix signatures and the stage timing.

The Agreement is a binding contract in itself and each signature is witnessed by two persons. As the Employer is one party to the Agreement, he will sign personally and not have the Architect sign on his behalf.

Note that Agreement NSC/2 becomes effective in stages and cross-references are made to the timing of signatures, etc. contained in Tender NSC/1 and Nomination NSC/3.

The first page of the Agreement NSC/2 will have the description of the Main Contract Works, Location and Sub-contract Works, exactly as stated in Tender NSC/1.

The date of the Agreement is particularly important and the sidenote clarifies the position as the date when the Architect signs Tender NSC/1 at Stage 5.

The date is followed by a full description of both the Employer's and the Sub-contractor's names and addresses. If either or both parties is a company or limited company, its full title should be entered.

Of the four recitals only the second requires the name of the Architect or Supervising Officer to be entered and it would be beneficial to delete whichever title does not apply. The recitals now effectively collect together the three documents NSC/1, NSC/2 and NSC/3 by stating that a tender has been submitted on Tender NSC/1. The Architect on the Employer's behalf has approved the Tender, and after agreement of Schedule 2 therein intends to nominate by NSC/3.

Clause 1   The Sub-contractor here is undertaking to make an agreement with the Contractor to settle the Particular Conditions of Schedule 2 of Tender NSC/1. He further undertakes to inform the Architect, through the Contractor, that agreement has been reached and the relevant page signed, which in turn will allow the Architect to issue Nomination NSC/3.

Clause 2.1   This is the Warranty referred to in the first page of explanatory notes to Tender NSC/1 and, as the clause indicates, covers warranty for design, selection of goods and materials and the satisfaction of performance specifications, all or any of which may form part of the Sub-contractor's Tender.

The Contractor is relieved of responsibility to the Employer in respect of Nominated Sub-contract works, with regard to the elements of the Warranty, in this clause of Agreement NSC/2. It has no significance whether the Agreement has been made or not.

Accordingly, the Employer requires the safeguard of this Warranty to give him direct legal rights against the Sub-contractor for failure or breach of the Warranty.

Clause 2.2   When at Stage 5 the Architect has signed Tender NSC/1 approving the Sub-contractor's Tender, he can issue an Instruction prior to Nomination, for the commencement of design work and ordering or fabrication of materials, directly to the Sub-contractor.

At this stage the Sub-contractor is still technically only 'proposed', as nomination has not yet taken place, and the clause covers payment to the Sub-contractor for this work, should the nomination not go ahead. The Employer would become the owner of the materials and have the use of the design work, but only so far as its use for the Sub-contract Works.

If nomination proceeds as expected, any payment made will be allowed in full against the amount due in respect of the Sub-contract Works.

Clause 3   This clause cannot operate or become an obligation upon the Sub-contractor until Nomination has taken place, and refers to the issue of drawings and information to the Architect in the proper time sequence.

The Main Contractor can claim an extension of time from the Employer if he does not receive information within a reasonable time from the Architect. If an extension of time is granted, that period is then obviously precluded to the Employer for the recovery of Liquidate and Ascertained Damages from the Contractor and he may further have to meet a claim for loss and expense.

If the delay in passing information by the Architect was caused initially by the Sub-contractor's late information to him, this clause of the Agreement gives the Employer the opportunity to recover his loss from the Sub-contractor, due to the granting of an extension of time to the Main Contractor.

The clause also refers to the performance of the Sub-contractor where he undertakes to relieve the Architect of the necessity of determining his employment.

If the Architect has to issue an Instruction for determination, the Employer then has the right and the opportunity to recover any loss he might sustain due to that determination.

Again, the Main Contractor can claim an extension of time from the Employer for delay on the part of a Nominated Sub-contractor which he has taken all reasonable steps to avoid. The result being that the Main Contractor is not responsible for the delay and therefore the Employer cannot claim Liquidated and Ascertained Damages. Under this clause of the Agreement the Employer's loss is now recoverable from the Sub-contractor.

Clause 21.8 Sub-contract NSC/4 states a valid suspension of the Sub-contract due to non-payment by the Contractor or the Employer, within the time scale. For the period of the suspension the Sub-contractor would be relieved of his obligations under this clause of the Agreement.

Lastly, the clause refers to the performance of the Sub-contractor which must be as he has tendered to perform and not involve the Architect in granting an extension of time to the Main Contractor. If, because of the Sub-contractor's performance, an extension of time is granted, then under this clause any loss and expense to the Employer caused by the extension is recoverable.

Clause 4   This is self-explanatory and fully amplified by the footnote. However, note that here the Employer is giving his agreement to an undertaking.

Clause 5.1   Again this is explained by the footnote, and again the Employer is assuming a liability.

Clause 5.2   This sub-clause states, that even if the final payment has been made to the Sub-contractor, he is still liable to rectify, at his own cost, any omission, fault or defect in his own works, for the full period up to the issue of the Final Certificate of the Main Contract. If he fails in this obligation, he will be liable to the Employer for all costs referred to in Clause 35.18 of the Main Contract Conditions.

Clause 5.3   The Sub-contractor, after the issue of the Final Certificate of the Main Contract, shall bear the same responsibility to the Main Contractor and the Employer for the Sub-contract Works as the Main Contractor does to the Employer for the Main Contract Works, all of which relates to the obligations of the contractor after the issue of the Final Certificate.

Clause 6   If the Sub-contractor defaults and causes the Architect to nominate another sub-contractor (Clause 35.24 of the Main Conditions) to complete, this clause gives the Employer the right to recover from the Sub-contractor all his costs or loss and expense due to renomination.

Note that if renomination becomes necessary due to determination by the Sub-contractor, then any additional costs are the liability of the Main Contractor (Clauses 35.24.3 and 35.24.6 of the Main Contract Conditions).

Clause 7   This clause is a further obligation undertaken by both the Architect and the Employer that the Architect will certify and the Employer will pay direct payments to the Sub-contractor under Clause 35.13.5 of the Main Conditions, if that clause operates.

This, however, is qualified by stating that any direct payments made after the start of a petition for winding up, or a resolution for winding up of the Main Contractor, will be repaid by the Sub-contractor to the Employer. This obviates the Employer from possibly paying twice, once to any Liquidator and also to the Sub-contractor (Clause 35.13.5.4.4 of the Main Conditions refers).

Clause 8   This clause operates only when Clause 2.3 of the Sub-contract Conditions is applicable and allows the Employer to enforce the Sub-contractor to honour his obligations to give the Contractor notice, with a copy to the Architect, of any restrictions, etc., as he becomes aware of them.

Clause 9   As the clause states, if there is any conflict between the two documents NSC/1 and NSC/2, then Agreement NSC/2 takes precedence.

Clause 10.1   Both parties to the Agreement are, by this clause, agreeing that disputes should go to Arbitration over anything arising out of the Agreement. Further, if they fail to agree a mutually satisfactory arbitrator, they agree to the appointment of a person as directed by the President or Vice-President of the Royal Institute of British Architects.

Clause 10.2 If the dispute which is being referred to arbitration under this Agreement is substantially the same as another dispute between

	(a)	Employer/Contractor	Main Contract,
or	(b)	Sub-Contractor/Contractor	NSC/4 or NSC/4a,
or	(c)	Employer/any other Sub-contractor	NSC/2 or NSC/2a,
or	(d)	Employer/Nominated Supplier	Clause 36.4.8, Main Contract,

and the dispute has already been referred to an Arbitrator, the Employer and Sub-contractor agree under this clause to accept the determination of the previously-appointed Arbitrator.

However, if either party considers that Arbitrator not appropriate to the current dispute, then a different Arbitrator may be appointed.

The last two paragraphs, which are Clauses 10.2.1 and 10.2.2, will not apply if the words 'Articles 5.1.4 and 5.1.5 apply' have been deleted from the Appendix to the Main Conditions.

Clause 10.3 Arbitration shall not commence until after
	(a)	Practical Completion of the Main Contract Works,
or	(b)	Alleged Practical Completion of the Main Contract Works
or	(c)	Termination of the Contractor's employment
or	(d)	Alleged termination of the Contractor's employment
or	(e)	Abandonment of the Main Contract Works
or	(f)	with the written consent of the Employer and the Sub-contractor

Clause 10.4 An arbitration award will be final and binding on the parties.

Clause 10.5 This clause is clearly explained by the footnote.

JCT

JCT Standard Form of Nomination of Sub-Contractor where Tender NSC/1 has been used

To: _____

(Main Contractor)

Main Contract Works: _____

_____ Job reference: _____

Sub-Contract Works:

ARCHITECT - STAGE 8

Page Number – Bills of Quantities or Specification: _____

Sub-Contractor hereby nominated: _____

of: _____

_____ Tel. No: _____

Further to my/our Preliminary Notice of Nomination (Main Contract Conditions clause 35·7·2)

dated   19

and Tender NSC/1 (and annexed documents) duly completed by you and the Sub-Contractor named above and by myself/ourselves on behalf of the Employer, the Sub-Contractor named above is hereby **NOMINATED** under the Main Contract Conditions clause 35·10·2 for the Sub-Contract Works identified above.

Signed _____ Architect/~~Supervising Officer~~

Address _____

_____

_____

Date   19

Circulation

☐ Main Contractor   ☐ Clerk of Works

☐ Quantity Surveyor   ☐ Consulting Engineer

☐ Sub-Contractor hereby nominated   ☐ Architect/Supervising Officer's file

© 1980 RIBA Publications Ltd

Explanatory notes

At Stage 8 on *Diagram 4* (page 122) the Nomination NSC/3 is sent by the Architect to the Contractor, with a copy to the Sub-contractor, and so forms part of the chain in the Basic Method of Nomination. In the Main Conditions Clause 35.10.2 directs that this procedure will be followed.

On the form the preliminary notice of nomination as required under Clause 35.7.2 is identified by its date, which the Architect will take care to ensure is correct. He will also check, prior to his nomination, that all the points to be agreed between the Contractor and the Sub-contractor in Tender NSC/1 have been so agreed and the deletions made and initialled.

Once the nomination has been issued, it leaves only the execution of NSC/4 between the Contractor and the Sub-contractor to complete the Basic Method.

1980 Edition

Nominated Sub-Contract NSC/4

JCT Standard Form of Sub-Contract for sub-contractors who have
tendered on Tender NSC/1 and executed Agreement NSC/2 and been
nominated by Nomination NSC/3 under the Standard Form of Building
Contract (clause 35.10.2) – Sub-Contract NSC/4

JCT
**Joint Contracts Tribunal for the
Standard Form of Building Contract**

156

50p stamp
to be
impressed here
if contract
is under seal.

Articles of Sub-Contract Agreement

made the _____ day of _____ 19____

Between _____

_____

of (or whose registered office is situated at)

_____

_____

(hereinafter called "the Contractor") of the one part and

CONTRACTOR - STAGE 9

OR ARCHITECT

of (or whose registered office is situated at)

_____

_____

(hereinafter called "the Sub-Contractor") of the other part.

Whereas

Recitals First the Sub-Contractor has submitted a tender on Tender NSC/1 for works (hereinafter called "the Sub-Contract Works") referred to on page 1 thereof and described in numbered documents annexed thereto and which are to be executed as part of works (hereinafter called "the Main Contract Works") referred to in Schedule 1 of the Tender and being carried out by the Contractor under a contract as described by or referred to in Schedule 1 of the Tender (hereinafter called "the Main Contract") with

(name of Employer) _____

_____

(hereinafter called "the Employer");

Second the tender on Tender NSC/1 has been duly completed and signed which Tender (comprising page 1 and the numbered documents annexed thereto and the completed Schedules 1 and 2 thereof) is hereinafter called "the Tender";

Third   the Contractor has retained the original of the Tender and the Sub-Contractor and the Employer each have a certified copy of the Tender and a further certified copy is appended hereto;

Fourth   the Sub-Contractor has been given notice of the provisions of the Main Contract as set out or referred to in Schedule 1 of the Tender except the detailed prices of the Contractor included in schedules and bills of quantities;

Fifth   the Contractor and the Sub-Contractor by a signature on page 1 of the Tender have agreed that the provisions of Sub-Contract NSC/4 shall apply unamended to the Sub-Contract for the Sub-Contract Works;

Sixth   the Architect has nominated the Sub-Contractor in Nomination NSC/3;

Seventh   at the date of the Sub-Contract:

CONTRACTOR
- STAGE 9

(A) the Sub-Contractor is/is not [a] the user of a current sub-contractor's tax certificate under the provisions of the Finance (No. 2) Act 1975 (hereinafter called "the Act") in one of the forms specified in Regulation 15 of the Income Tax (Sub-Contractors in the Construction Industry) Regulations, 1975 and the Schedule thereto (hereinafter called "the Regulations");

(B) the Contractor is/is not [a] the user of a current sub-contractor's tax certificate under the Act and the Regulations;

(C) the Employer under the Main Contract is/is not [a] a "contractor" within the meaning of the Act and the Regulations.

Now it is hereby agreed as follows

Sub-Contractor's obligation   1·1

Article 1
In accordance with the Tender the Sub-Contractor will carry out and complete the Sub-Contract Works.

Sub-Contract Sum   2·1

Article 2
Where a Sub-Contract Sum is stated on page 1 of the Tender the Contractor will pay to the Sub-Contractor that Sub-Contract Sum or such other sum as shall become payable in accordance with the Sub-Contract.

Ascertained Final Sub-Contract Sum   2·2

Where a Tender Sum is stated on page 1 of the Tender the Contractor will pay to the Sub-Contractor such sum or sums as shall become payable in accordance with the Sub-Contract and the total of such sums is in Sub-Contract NSC/4 called "the Ascertained Final Sub-Contract Sum".

Footnote

[a] Delete whichever alternative is not applicable; such deletion must follow the way in which the Tender, Schedule 1, item 10 and Schedule 2, item 7 have been completed unless the position as set out in the Tender has changed by the time NSC/4 is executed; in such circumstances the Seventh recital should be correctly completed and the Tender corrected so that the Seventh recital and the Tender are not in conflict

Article 3

Settlement of
disputes -
arbitration

3·1 In the event of any dispute or difference between the Contractor and Sub-Contractor, whether arising during the execution or after the completion or abandonment of the Sub-Contract Works or after the determination of the employment of the Sub-Contractor under Sub-Contract NSC/4 (whether by breach or in any other manner), in regard to any matter or thing of whatsoever nature arising out of the Sub-Contract or in connection therewith, then either party shall give to the other notice in writing of such dispute or difference and such dispute or difference shall be and is hereby referred to the arbitration and final decision of a person to be agreed between the parties, or, failing such agreement within 14 days after either party has given to the other a written request to concur in the appointment of an Arbitrator, a person to be appointed on the request of either party by the President or a Vice-President for the time being of the Royal Institution of Chartered Surveyors or, alternatively, if the party requiring the appointment so decides, by the President or a Vice-President for the time being of the Royal Institute of British Architects.

3·2 ·1 Provided that if the dispute or difference to be referred to arbitration under this Sub-Contract raises issues which are substantially the same as or connected with issues raised in a related dispute between

the Contractor and the Employer under the Main Contract or

the Sub-Contractor and the Employer under Agreement NSC/2 or NSC/2a as applicable or

the Contractor and any other nominated sub-contractor under Sub-Contract NSC/4 or NSC/4a as applicable or

the Contractor and any Nominated Supplier whose contract of sale with the Contractor provides for the matters referred to in clause 36·4·8 of the Main Contract Conditions

and if the related dispute has already been referred for determination to an Arbitrator, the Contractor and the Sub-Contractor hereby agree that the dispute or difference under this Sub-Contract shall be referred to the Arbitrator appointed to determine the related dispute; and such Arbitrator shall have power to make such directions and all necessary awards in the same way as if the procedure of the High Court as to joining one or more defendants or joining co-defendants or third parties was available to the parties and to him.

·2 Save that the Contractor or the Sub-Contractor may require the dispute or difference under this Sub-Contract to be referred to a different Arbitrator (to be appointed under this Sub-Contract) if either of them reasonably considers that the Arbitrator appointed to determine the related dispute is not appropriately qualified to determine the dispute or difference under this Sub-Contract.

·3 Articles 3·2·1 and 3·2·2 shall apply unless in the Tender, Schedule 1, item 10 the words "Articles 5·1·4 and 5·1·5 apply" have been deleted.

3·3 Such Arbitrator shall not without the written consent of the Contractor and Sub-Contractor enter on the arbitration until after the practical completion or abandonment of the Main Contract Works, except to arbitrate:

3·3 ·1 whether a payment has been improperly withheld or is not in accordance with the Sub-Contract; or

3·3 ·2 whether practical completion of the Sub-Contract Works shall be deemed to have taken place under clause 14·2; or

3·3 ·3 in respect of a claim by the Contractor or counterclaim by the Sub-Contractor to which the provisions of clause 24 apply in which case the Arbitrator shall exercise the powers given to him in clause 24; or

3·3 ·4 any matters in dispute under clause 4·3 in regard to reasonable objection by the Sub-Contractor or under clauses 11·2 and 11·3 as to extension of time.

3·4 In any such arbitration as is provided for in article 3 any decision of the Architect which is final and binding on the Contractor under the Main Contract shall also be and be deemed to be final and binding between and upon the Contractor and the Sub-Contractor.

3·5   Subject to the provisions of clauses 4·6, 35·4·3, 36·5·3, 37·5 and clause 30 of the Main Contract Conditions the Arbitrator shall, without prejudice to the generality of his powers, have power to direct such measurements and/or valuations as may in his opinion be desirable in order to determine the rights of the parties and to ascertain and award any sum which ought to have been the subject of or included in any certificate and to open up, review and revise any certificate, opinion, decision, requirement or notice and to determine all matters in dispute which shall be submitted to him in the same manner as if no such certificate, opinion, decision, requirement or notice had been given.

3·6   The award of such Arbitrator shall be final and binding on the parties.

3·7   Whatever the nationality, residence or domicile of the Employer, the Contractor, the Sub-Contractor, and any sub-contractor or supplier or the Arbitrator, and wherever the Main Contract Works or Sub-Contract Works, or any parts thereof, are situated, the law of England shall be the proper law applicable to the Sub-Contract and in particular (but not so as to derogate from the generality of the foregoing) the provisions of the Arbitration Acts, 1950 (notwithstanding anything in S.34 thereof) to 1979 shall apply to any arbitration under article 3 wherever the same, or any part of it, shall be conducted. **[b]**

Attestation   (A)   Signed by or on behalf of the Contractor in the presence of: **[c]**

CONTRACTOR - STAGE 9

Signed by or on behalf of the Sub-Contractor in the presence of: **[c]**

NOMINATED SUB-CONTRACTOR - STAGE 10

(B)   The Common Seal of the Contractor was hereunto affixed in the presence of: **[d]**

(C.S.)

The Common Seal of the Sub-Contractor was hereunto affixed in the presence of: **[d]**

(C.S.)

Footnotes

[b] Where the parties do not wish the proper law of the Sub-Contract to be the law of England and/or do not wish the provisions of the Arbitration Acts 1950 to 1979 to apply to any arbitration under the Sub-Contract held under the procedural law of Scotland (or other country) appropriate amendments to article 3·5 should be made.

[c] Delete if Tender, Schedule 2, item 11 states that Sub-Contract NSC/4 is to be executed under seal and execute as provided in (B).

[d] Delete if Tender, Schedule 2, item 11 states that Sub-Contract NSC/4 is to be executed under hand and execute as provided in (A).

Explanatory notes

The Nominated Sub-contract NSC/4 is reproduced here only as far as the Articles of Agreement and Attestation. Following these documents are the Sub-contract Conditions annexed to the Articles of Sub-contract Agreement. The Sub-contract Conditions are the subject of a companion book in this series and discussed in detail therein.

There is no Appendix to this Contract as the information in Tender NSC/1 is detached and attached to the end of the Contract Document.

Nomination procedure – Alternative Method

The Main Conditions Clauses 35.11 and 35.12 govern the procedures for nomination under the Alternative Method.

This is the shorter of the two methods used to obtain a nomination but it has some disadvantages. In the Basic Method an agreement and contract is formed between all the parties, Architect (Employer), Contractor and proposed Nominated Sub-contractor. In the Alternative Method the same holds good only if Agreement NSC/2a is used. If it is not then there is no agreement between the Employer and the Sub-contractor.

In Tender NSC/1 of the Basic Method, the assurance is made that all matters relating to programme and the integration of the Nominated Sub-contract Works into the Main Contractor's arrangements for the works as a whole, are agreed before nomination takes place. Under the Alternative Method this assurance is not made, as there is no tender document, and these matters may have to be settled after the nomination has been issued.

If Agreement NSC/2a is not used, all the advantages with regard to direct contractual rights for both parties are lost.

The Alternative Method can still be adopted whether it be before or after the Main Contract is let.

For Clauses 35.11 and 35.12 to operate contractually, the following requirements must be met:

A statement to be made that Clauses 35.11 and 35.12 will apply to a proposed Nominated Sub-contract in the following circumstances:

1 Naming a Nominated Sub-Contract in the Contract Bills (Clause 35.5.1.2), OR
2 A Prime Cost Sum in the Main Contract Bills (Clause 35.5.1.2), OR
3 An Architect's Instruction requiring a variation which nominates a Sub-contractor (Clause 35.5.1.2) check also Clause 35.1.3, OR
4 An Architect's Instruction expending a Provisional Sum in the Main Contract which nominates a Sub-contractor (Clause 35.5.1.2).

Note: If no such statement is made in any of the above circumstances, the Basic Method will operate.

Several other criteria must also be met, as follows:

5 The statement in any of the circumstances 1 to 4 above must also state whether or not the Sub-contractor has to execute Agreement NSC/2a (Clause 35.5.1.2).

6 If required, Agreement NSC/2a must be executed by the Employer and the Sub-contractor (Clause 35.11.1).

7 When the Architect issues his nomination in an Instruction, a copy must go to the Sub-contractor (Clause 35.11.2).

8 The Contractor has not made a reasonable objection to the proposed Sub-contractor within seven days of receipt of the Instruction (Clause 35.4.3).

9 The Contractor and Sub-contractor to execute NSC/4a within 14 days of the Instruction. Also the Contractor to inform the Architect immediately if NSC/4a is not executed within that period (Clause 35.12).

The Alternative Method has no tender document, and Tender NSC/1 must not be used for this purpose. There is, therefore, a problem in the transmission of the information regarding the terms of the Main Contract and the terms of the Sub-contract. It is essential that tenderers who are to be nominated have received the complete terms of the Sub-contract, which must include all the information necessary to allow the completion of the Appendix to Sub-contract NSC/4a. The information required in the Appendix is very similar to that of Tender NSC/1 in the Basic Method.

Some form of tender must be submitted by the proposed Sub-contractor, and if the information given to him at the time of tender is insufficient then he should add the information in his Tender.

Diagram 8, next page, shows the complete procedure from start to finish, all of which is covered in Clauses 35.11 and 35.12 of the Conditions.

Diagram 9 shows the procedure if the proposed Sub-contractor fails to enter into a contract without due cause (Clause 35.23.3), while *Diagram 10* shows the procedure if the Contractor make a reasonable objection within seven days (Clause 35.4.3).

Immediately following *Diagrams 8, 9* and *10* are the forms indicating where the various parties have to provide information and/or signatures to effect a nomination, together with explanatory notes.

The timing of the giving of information and/or signatures is important, and with this in view, the various Stage numbers to accord with *Diagram 8* and name of party, are also entered in the forms.

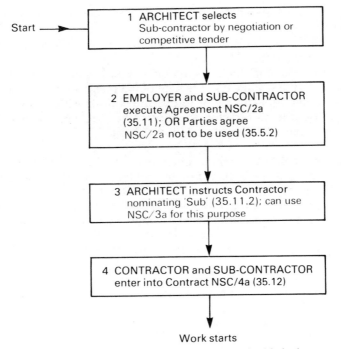

Diagram 8: Nomination of a Sub-contractor using the Alternative Method

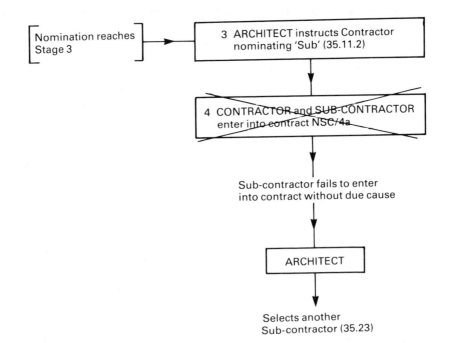

Diagram 9: Nomination of a Sub-contractor using the Alternative Method – Proposed Sub-contractor fails to enter into contract without due cause, 35.23.3

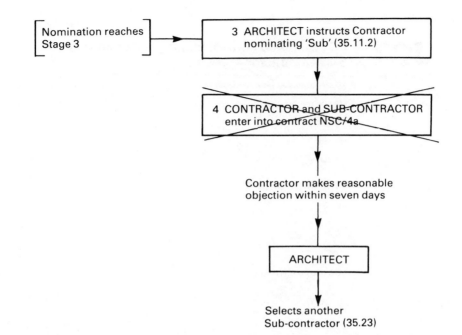

Diagram 10: Nomination of a Sub-contractor using the Alternative Method – Contractor makes reasonable objection within seven days, 35.4.3

JCT

JCT Standard Form of Employer/Nominated Sub-Contractor Agreement

Agreement between a Sub-Contractor to be nominated for Sub-Contract Work in accordance with clauses 35·11 and 35·12 of the Standard Form of Building Contract for a main contract and the Employer referred to in the main contract.

Main Contract Works:

Location:

Sub-Contract Works:

ARCHITECT - STAGE 2

This Agreement

The date to be inserted here must be a date not later than the date of the nomination instruction of the Architect/Supervising Officer under clause 35·11 of the Main Contract Conditions

is made the _____ day of _____ 19 _____

between _____

of or whose registered office is situated at _____

_____

(hereinafter called 'the Employer') and

_____

of or whose registered office is situated at _____

_____

(hereinafter called 'the Sub-Contractor')

Whereas

Identify edition and revision, if any, of the Standard Form

First the Employer has entered into/intends to enter into* a contract (hereinafter called 'the Main Contract') for certain works (hereinafter called 'the Main Contract Works') and for which the Main Contract Conditions are:

_____

(hereinafter called 'the Main Contract Conditions');

Second the Employer has appointed

_____

to be the Architect/Supervising Officer for the purposes of the Main Contract and this Agreement (hereinafter called 'the Architect/Supervising Officer' which expression as used in this Agreement shall include his successor validly appointed under the Main Contract or otherwise before the Main Contract is operative).

Third the Sub-Contractor has tendered for certain works (hereinafter called 'the Sub-Contract Works') to be carried out by a Nominated Sub-Contractor as part of the Main Contract Works and the tender of the Sub-Contractor has been made on the basis that the Sub-Contractor and the Employer will enter into Agreement NSC/2a;

Fourth the Architect/Supervising Officer on behalf of the Employer has approved the tender of the Sub-Contractor and the Main Contract (either in the Contract Bills forming part of the Main Contract or in an instruction of the Architect/Supervising Officer under the Main Contract Conditions) provides that clauses 35·11 and 35·12 of the Main Contract Conditions shall apply to the nomination of the Sub-Contractor;

Fifth the Architect/Supervising Officer has issued/intends to issue* an instruction under clause 35·11 of the Main Contract Conditions nominating the Sub-Contractor to carry out and complete the Sub-Contract Works on the terms and conditions of Sub-Contract NSC/4a as referred to in clause 35·12 of the Main Contract Conditions;

Sixth nothing contained in this Agreement is intended to render the Architect/Supervising Officer in any way liable to the Sub-Contractor in relation to matters in the said Agreement.

Now it is hereby agreed

Design, materials,
performance
specification

1·1   The Sub-Contractor warrants that he has exercised and will exercise all reasonable skill and care in

·1   the design of the Sub-Contract Works insofar as the Sub-Contract Works have been or will be designed by the Sub-Contractor; and

·2   the selection of materials and goods for the Sub-Contract Works insofar as such materials and goods have been or will be selected by the Sub-Contractor; and

·3   the satisfaction of any performance specification or requirement insofar as such performance specification or requirement is included or referred to in the Sub-Contract Documents as defined in Sub-Contract NSC/4a.

Nothing in clause 1·1 shall be construed so as to affect the obligations of the Sub-Contractor under Sub-Contract NSC/4a in regard to the supply under the Sub-Contract of workmanship, materials and goods.

1·2   ·1   If, after the date of this Agreement and before the issue by the Architect/Supervising Officer of the instruction under clause 35·11 of the Main Contract Conditions, the Architect/Supervising Officer instructs in writing that the Sub-Contractor should proceed with
·1   the designing of, or
·2   the proper ordering or fabrication of any materials or goods for
the Sub-Contract Works the Sub-Contractor shall forthwith comply with the instruction and the Employer shall make payment for such compliance in accordance with clauses 1·2·2 to 1·2·4.

·2   No payment referred to in clauses 1·2·3 and 1·2·4 shall be made after the issue of the instruction nominating the Sub-Contractor under clause 35·11 of the Main Contract Conditions except in respect of any design work properly carried out and/or materials or goods properly ordered or fabricated in compliance with an instruction under clause 1·2·1 but which are not used for the Sub-Contract Works by reason of some written decision against such use given by the Architect/Supervising Officer before the issue of the instruction nominating the Sub-Contractor under clause 35·11 of the Main Contract Conditions.

·3   The Employer shall pay the Sub-Contractor the amount of any expense reasonably and properly incurred by the Sub-Contractor in carrying out work in the designing of the Sub-Contract Works and upon such payment the Employer may use that work for the purposes of the Sub-Contract Works but not further or otherwise.

·4   The Employer shall pay the Sub-Contractor for any materials or goods properly ordered by the Sub-Contractor for the Sub-Contract Works and upon such payment any materials and goods so paid for shall become the property of the Employer.

·5   If any payment has been made by the Employer under clauses 1·2·3 and 1·2·4 and the Sub-Contractor is subsequently nominated by an instruction nominating the Sub-Contractor issued under clause 35·11 of the Main Contract Conditions to execute the Sub-Contract Works the Sub-Contractor shall allow to the Employer and the Main Contractor full credit for such payment in the discharge of the amount due in respect of the Sub-Contract Works.

Delay in supply of
information and in
performance by
Sub-Contractor

2·1   The Sub-Contractor will not be liable under clauses 2·2, 2·3 or 2·4 until the Architect/Supervising Officer has issued his instruction nominating the Sub-Contractor issued under clause 35·11 of the Main Contract Conditions nor in respect of any revised period of time for delay in carrying out or completing the Sub-Contract Works which the Sub-Contractor has been granted under clause 11·2 of Sub-Contract NSC/4a.

2·2   The Sub-Contractor shall so supply the Architect/Supervising Officer with information (including drawings) in accordance with the agreed programme details or at such time as the Architect/Supervising Officer may reasonably require so that the Architect/Supervising Officer will not be delayed in issuing necessary instructions or drawings under the Main Contract, for which delay the Main Contractor may have a valid claim to an extension of time for completion of the Main Contract Works by reason of the Relevant Event in clause 25·4·6 or a valid claim for direct loss and/or expense under clause 26·2·1 of the Main Contract Conditions.

2·3   The Sub-Contractor shall so perform his obligations under the Sub-Contract that the Architect/Supervising Officer will not by reason of any default by the Sub-Contractor be under a duty to issue an instruction to determine the employment of the Sub-Contractor under clause 35·24 of the Main Contract Conditions provided that any suspension by the Sub-Contractor of further execution of the Sub-Contract Works under clause 21·8 of Sub-Contract NSC/4a shall not be regarded as a 'default by the Sub-Contractor' as referred to in clause 2·3.

2·4   The Sub-Contractor shall so perform the Sub-Contract that the Contractor will not become entitled to an extension of time for completion of the Main Contract Works by reason of the Relevant Event in clause 25·4·7 of the Main Contract Conditions.

166

Architect's direction on value of Sub-Contract Work in Interim Certificates – information to Sub-Contractor	3	The Architect/Supervising Officer shall operate the provisions of *clause 35·13·1 of the Main Contract Conditions.
Employer's duty – final payment for Sub-Contract Works	4·1	The Architect/Supervising Officer shall operate the provisions in †clauses 35·17 to 35·19 of the Main Contract Conditions.
Discharge of final payment to Sub-Contractor – Sub-Contractor's obligations	4·2	After due discharge by the Contractor of a final payment under clause 35·17 of the Main Contract Conditions the Sub-Contractor shall rectify at his own cost (or if he fails so to rectify, shall be liable to the Employer for the costs referred to in clause 35·18 of the Main Contract Conditions) any omission, fault or defect in the Sub-Contract Works which the Sub-Contractor is bound to rectify under Sub-Contract NSC/4a after written notification thereof by the Architect/Supervising Officer at any time before the issue of the Final Certificate under clause 30·8 of the Main Contract Conditions.

4·3 After the issue of the Final Certificate under the Main Contract Conditions the Sub-Contractor shall in addition to such other responsibilities, if any, as he has under this Agreement, have the like responsibility to the Main Contractor and to the Employer for the Sub-Contract Works as the Main Contractor has to the Employer under the terms of the Main Contract relating to the obligations of the Contractor after the issue of the Final Certificate.

Architect's instructions – duty to make a further nomination – liability of Sub-Contractor	5	Where the Architect/Supervising Officer has been under a duty under clause 35·24 of the Main Contract Conditions except as a result of the operation of clause 35·24·6 to issue an instruction to the Main Contractor making a further nomination in respect of the Sub-Contract Works, the Sub-Contractor shall indemnify the Employer against any direct loss and/or expense resulting from the exercise by the Architect/Supervising Officer of that duty.
Architect's certificate of non-discharge by Contractor – payment to Sub-Contractor by Employer	6·1	The Architect/Supervising Officer and the Employer shall operate the provisions in regard to the payment of the Sub-Contractor in clause 35·13 of the Main Contract Conditions.

6·2 If, after paying any amount to the Sub-Contractor under clause 35·13·5·3 of the Main Contract Conditions, the Employer produces reasonable proof that there was in existence at the time of such payment a petition or resolution to which clause 35·13·5·4·4 of the Main Contract Conditions refers, the Sub-Contractor shall repay on demand such amount.

Conditions in contracts for purchase of goods and materials – Sub-Contractor's duty	7	Where ‡clause 2·3 of Sub-Contract NSC/4a applies, the Sub-Contractor shall forthwith supply to the Contractor details of any restriction, limitation or exclusion to which that clause refers as soon as such details are known to the Sub-Contractor.
Arbitration	8·1	In case any dispute or difference shall arise between the Employer or the Architect/Supervising Officer on his behalf and the Sub-Contractor, either during the progress or after the completion or abandonment of the Sub-Contract Works, as to the construction of this Agreement, or as to any matter or thing of whatsoever nature arising out of this Agreement or in connection therewith, then such dispute or difference shall be and is hereby referred to the arbitration and final decision of a person to be agreed between the parties, or, failing agreement within 14 days after either party has given to the other a written request to concur in the appointment of an Arbitrator, a person to be appointed on the request of either party by the President or a Vice President for the time being of the Royal Institute of British Architects.

8·2 ·1 Provided that if the dispute or difference to be referred to arbitration under this Agreement raises issues which are substantially the same as or connected with issues raised in a related dispute between the Employer and the Contractor under the Main Contract or between the Sub-Contractor and the Contractor under Sub-Contract NSC/4 or NSC/4a or between the Employer and any other nominated Sub-Contractor under Agreement NSC/2 or NSC/2a or between the Employer and any Nominated Supplier whose contract of sale with the Main Contractor provides for the matters referred to in clause 36·4·8 of the Main Contract Conditions, and if the related dispute has already been referred for determination to an Arbitrator, the Employer and the Sub-Contractor hereby agree that the dispute or difference under this Agreement shall be referred to the Arbitrator appointed to determine the related dispute; and such Arbitrator shall have power to make such directions and all necessary awards in the same way as if the procedure of the High Court as to joining one or more defendants or joining co-defendants or third parties was available to the parties and to him.

8·2 ·2 Save that the Employer or the Sub-Contractor may require the dispute or difference under this Agreement to be referred to a different Arbitrator (to be appointed under this Agreement) if either of them reasonably considers that the Arbitrator appointed to determine the related dispute is not appropriately qualified to determine the dispute or difference under this Agreement.

---

*Note: Clause 35·13·1 requires that the Architect/Supervising Officer, upon directing the Main Contractor as to the amount included in any Interim Certificates in respect of the value of the Nominated Sub-Contract Works issued under clause 30 of the Main Contract Conditions, shall forthwith inform the Sub-Contractor in writing of that amount.

†Note: Clause 35·17 deals with final payment by the Employer for the Sub-Contract Works prior to the issue of the Final Certificate under the Main Contract Conditions.

‡Note: Clause 2·3 deals with specified supplies and restrictions etc. in the contracts of sale for such supplies.

Agreement NSC/2a

8·2   ·3   Clauses 8·2·1 and 8·2·2 shall apply unless in the Appendix to the Main Contract Conditions the words 'Articles 5·1·4 and 5·1·5 apply' have been deleted.

8·3   Such reference shall not be opened until after Practical Completion or alleged Practical Completion of the Main Contract Works or termination or alleged termination of the Contractor's employment under the Main Contract or abandonment of the Main Contract Works, unless with the written consent of the Employer or the Architect/Supervising Officer on his behalf and the Sub-Contractor.

8·4   The award of such Arbitrator shall be final and binding on the parties.

8·5*   Whatever the nationality, residence or domicile of the Employer, the Contractor, any Sub-Contractor or supplier or the Arbitrator, and wherever the Works or any part thereof are situated, the law of England shall be the proper law of this Agreement and in particular (but not so as to derogate from the generality of the foregoing) the provisions of the Arbitration Acts 1950 (notwithstanding anything in S·34 thereof) to 1979 shall apply to any arbitration under Clause 8 wherever the same, or any part of it, shall be conducted.

*Where the parties do not wish the proper law of the contract to be the law of England and/or do not wish the provisions of the Arbitration Acts 1950 to 1979 to apply to any arbitration under the Contract held under the procedural law of Scotland (or other country) appropriate amendments to Clause 8·5 should be made.

Signed by or on behalf of the Sub-Contractor [g1] ____ SUB-CONTRACTOR - STAGE 2

in the presence of:

Signed by or on behalf of the Employer [g1] ____ EMPLOYER - STAGE 2

in the presence of:

Signed, sealed and delivered by [g2]/The common seal of [g3]: ____

in the presence of [g2]/was hereunto affixed in the presence of [g3]:

Signed, sealed and delivered by [g2]/The common seal of [g3]: ____

in the presence of [g2]/was hereunto affixed in the presence of [g3]:

Footnotes   [g1] For use if Agreement is executed under hand.   [g2] For use if executed under seal by an individual or firm or unincorporated body.   [g3] For use if executed under seal by a company or other body corporate.

Page 4

Explanatory notes

As already explained, under the Alternative Method this document may or may not be used, according to the decision of the Employer. Whether or not it is to be used in the nomination procedure must be stated both to the Contractor and the proposed Sub-contractor. Any one of the four methods by which the Architect intimates his intention to make a nomination will additionally include this statement. The Sub-contractor will be informed at the time of his being invited to tender.

It may well be that if NSC/2a is to be used it could be agreed and signed well before nomination takes place so as to allow design and ordering of specialist materials to be put well in hand.

Under this method of nomination there is no printed tender document and the proposed Sub-contractor will tender in his own fashion. The Tender has to be approved by the Architect, which can be done by letter to the Employer and the Sub-contractor, or by merely counter-signing the Tender, stating that it is approved, with the date of the approval. Some Architects may have a 'stamp' for this purpose.

As the Agreement is a binding contract between the Employer and the Sub-contractor, the two signatures are witnessed, and for the same reason the Architect cannot sign on behalf of his Employer.

Again the form has its parts indicated to show where the parties enter the information or affix signatures and the stage timing.

The two forms NSC/2 and NSC/2a are broadly similar, but some further detailed notes are necessary.

The first page of the Agreement NSC/2a will have the description of the Main Contract Works, location and Sub-contract Works as filled in for the Agreement in the Basic Method. The date is again clarified by the sidenote as being a date not later when the Architect issues his nomination instruction.

The first recital will state, by deletion, whether or not the Main Contract has been let and under which Conditions the Main Contract will be governed.

The second recital again states the name of the Architect or Supervising Officer and the non-applicable title will be struck out.

The third and fourth recitals state that the Sub-contractor has tendered and on the basis of NSC/2a being used. That the Tender has been approved and that Clauses 35.11 and 35.12 will apply to the nomination.

The fifth recital, by deletion, states whether or not an Instruction for nomination has been issued.

The sixth recital is the same as the fourth in NSC/2 with, of course, the words 'and Tender' deleted.

Clause 1.1   The Sub-contractor here gives a design warranty for any design work, selection of materials and goods and the satisfaction of any performance specification required, all or any of which, may form part of the Sub-contractor's Tender.

The Contractor is relieved of responsibility to the Employer under Clause 35.21 in respect of Nominated Sub-contract Works with regard to the elements of the Warranty in this Clause of Agreement NSC/2a. It has no significance whether the Agreement has been made or not.

Accordingly, the Employer requires the safeguard of this Warranty to give him direct legal rights against the Sub-contractor in the event of failure or breach of the Warranty.

This clause becomes effective when the Architect approves the Tender, thus allowing the Architect to instruct in advance of nomination the commencement of design work and the ordering or fabricating of materials.

If no design work is necessary by the Sub-contractor, then the clause does not operate. If design work is necessary, it must be fully described in the description of the Sub-Contract Works given in the Tender.

Clause 1.2 When the Architect has approved the Sub-contractor's Tender, he can issue an Instruction prior to nomination, for the commencement of design work and ordering or fabrication of materials, directly to the Sub-contractor.

At this stage since nomination has not yet taken place, the Sub-contractor is still technically only 'proposed' and this clause covers for payment to the Sub-contractor for this work, should the nomination not go ahead. The Employer would become the owner of the materials and have the use of the design work, but only so far as its use for the Sub-contract Works.

If nomination proceeds as expected, any payment made, will be allowed in full, against the amount due in respect of the Sub-contract Works.

Clause 2 This clause cannot operate or become an obligation upon the Sub-contractor until nomination has taken place. Nor is the Sub-contractor liable in respect of a revised period of time for delay granted under Clause 11.2 of the Sub-contract NSC/4a.

If a delay in passing information to the Architect was caused initially by the Sub-contractor's late information to him, this clause gives the Employer the opportunity to recover his loss from the Sub-contractor, due to the granting of an extension of time to the Main Contractor.

The clause also refers to the performance of the Sub-contractor where he undertakes to relieve the Architect of the necessity of determining his employment. If the Architect has to issue an Instruction for Determination, the Employer has the right and opportunity to recover any loss he might sustain due to that determination.

The Main Contractor can claim an extension of time from the Employer for delay on the part of a Nominated Sub-contractor which he has taken all reasonable steps to avoid. The result being that the Main Contractor is not responsible for the delay and therefore the Employer cannot claim Liquidated and Ascertained Damages. The Employer's loss is now recoverable from the Sub-contractor under this Agreement.

Clause 21.8, Sub-contract NSC/4a states a valid suspension of the Sub-contract due to non-payment by the Contractor, or the Employer, within the time scale. For the period of the suspension, the Sub-Contractor would be relieved of his obligation under this clause of the Agreement.

Lastly, the Sub-contractor's performance of his Sub-contract will be such as not to entitle the Contractor to an extension of time under the

Main Contract (Clause 25.4.7). If an extension is granted, the Employer can recover from the Sub-contractor any loss and expense caused to him.

Clause 3   As in Agreement NSC/2 the Employer is giving his agreement to an undertaking, which is explained in the footnote.

Clause 4.1   Again the Employer is assuming a liability as explained in the footnote.

Clause 4.2   This sub-clause states, that even if the final payment has been made to the Sub-contractor, he is still liable to rectify, at his own cost, any omission, faults or defects in his own works, for the full period up to the issue of the Final Certificate of the Main Contract. If he fails in this obligation, he will be liable to the Employer for all costs referred to in Clause 35.18 of the Main Contract Conditions.

Clause 4.3   After the issue of the Final Certificate of the Main Contract, the Sub-contractor, shall bear the same responsibility to the Contractor and the Employer for the Sub-contract Works as the Contractor does to the Employer for the Main Contract Works, all of which relates to the obligations of the Contractor after the issue of the Final Certificate.

Clause 5   If the Sub-contractor defaults and causes the Architect to nominate another Sub-Contractor (Clause 35.24 of the Main Contract Conditions) to complete, the Employer is given the right to recover all his costs or loss and expenses due to the renomination, from the Sub-Contractor.

Note that if renomination becomes necessary due to determination by the Sub-contractor then, in that case, any additional costs are the liability of the Main Contractor (Clauses 35.24.3 and 35.24.6).

Clause 6   Here a further obligation is undertaken by the Architect to certify, and the Employer to pay, direct payments to the Sub-contractor under Clause 35.13.5 of the Main Contract Conditions, if that clause operates.

This is qualified by stating that any direct payments made after the start of a petition for winding up, or a resolution for winding up the Main Contractor, will be repaid by the Sub-contractor to the Employer. This obviates the Employer from possibly paying twice, once to any liquidator and also to the Sub-contractor (Clause 35.13.5.4.4 of the Main Contract Conditions refers).

Clause 7   This Clause operates only when Clause 2.3 of the Sub-contract Conditions is applicable, and allows the Employer to force the Sub-contractor to honour his obligations to give the Contractor notice, with a copy to the Architect, of any restrictions, etc. as he becomes aware of them.

Clause 8.1 to Clause 8.5   These Clauses have exactly the same wording as Clauses 10.1 to 10.5 of Agreement NSC/2 with reference to arbitration and do not need to be reiterated here.

1980 Edition

Nominated Sub-Contract NSC/4a

JCT Standard Form of Sub-Contract for sub-contractors nominated
under the Standard Form of Building Contract (clauses 35·11 and 35·12)

JCT
**Joint Contracts Tribunal for the
Standard Form of Building Contract**

172

50p stamp
to be
impressed here
if contract
is under seal.

Articles of Sub-Contract Agreement

made the _____ day of _____ 19___

Between _____

_____

of (or whose registered office is situated at)

_____

CONTRACTOR - STAGE 1~~
OR ARCHITECT~~

(hereinafter called "the Contractor") of the one part and

_____

_____

of (or whose registered office is situated at)

_____

_____

(hereinafter called "the Sub-Contractor") of the other part.

Whereas

Recitals First the Sub-Contractor has submitted an offer to carry out works (hereinafter called "the Sub-Contract Works") referred to in the Appendix, part 1, and described in numbered documents identified in that part of the Appendix (hereinafter called the "Numbered Documents");

Second the Sub-Contract Works are to be executed as part of works (hereinafter called "the Main Contract Works") referred to in the Appendix, part 2 Section A and being carried out by the Contractor under a contract as described in the Appendix, part 2, (hereinafter called "the Main Contract") with

_____
(name of Employer)

hereinafter called "the Employer");

Third pursuant to clause 35 of the Main Contract Conditions the Architect has selected and approved the Sub-Contractor to carry out the Sub-Contract Works/* and the Employer and the Sub-Contractor have entered into Agreement NSC/2a;

Delete if the Employer and the Sub-Contractor have agreed not to enter into Agreement NSC/2a.

Fourth   under clause 35·11 of the Main Contract Conditions the Architect has issued an instruction (witha copy to the Sub-Contractor) dated _____ nominating the Sub-Contractor to carry out the Sub-Contract Works and the Sub-Contractor has tendered on the basis that he will enter into Sub-Contract NSC/4a with the Contractor within 14 days of the date of the nomination instruction.

CONTRACTOR - STAGE 4

Fifth   at the date of the Sub-Contract:

(A) the Sub-Contractor is/is not [a] the user of a current sub-contractor's tax certificate under the provisions of the Finance (No. 2) Act 1975 (hereinafter called "the Act") in one of the forms specified in Regulation 15 of the Income Tax (Sub-Contractors in the Construction Industry) Regulations, 1975 and the Schedule thereto (hereinafter called "the Regulations");

CONTRACTOR - STAGE 4

(B) the Contractor is/is not [a] the user of a current sub-contractor's tax certificate under the Act and the Regulations;

(C) the Employer under the Main Contract is/is not [a] a "contractor" within the meaning of the Act and the Regulations.

Now it is hereby agreed as follows

Sub-Contractor's obligation

Article 1

1·1   The Sub-Contractor will upon and subject to Sub-Contract NSC/4a carry out and complete the Sub-Contract Works shown upon and described by or referred to in the Numbered Documents and in Sub-Contract NSC/4a.

Sub-Contract Sum

Article 2

2·1*   The Contractor will pay to the Sub-Contractor the VAT-exclusive sum of

£ _____ *CONTRACTOR - STAGE 4* _____

_____ (words)

which sum takes into account the 2½ per cent cash discount allowable to the Contractor under Sub-Contract NSC/4a (hereinafter referred to as "the Sub-Contract Sum") or such other sum as shall become payable in accordance with the Sub-Contract.

Ascertained Final Sub-Contract Sum

2·2*   The VAT-exclusive Tender Sum is

£ _____ *CONTRACTOR - STAGE 4* _____

_____ (words)

which sum takes into account the 2½ per cent cash discount allowable to the Contractor under Sub-Contract NSC/4a (hereinafter referred to as "the Tender Sum") and the Contractor will pay to the Sub-Contractor such sum or sums as shall become payable in accordance with the Sub-Contract and the total of such sums is in the Sub-Contract called "the Ascertained Final Sub-Contract Sum".

Delete article 2·1 or 2·2 as appropriate.

Footnote   [a] Delete whichever alternative is not applicable; such deletion must follow the way in which the Appendix, part 2 Section B and part 12 have been completed.

Article 3

Settlement of
disputes –
arbitration

3.1 In the event of any dispute or difference between the Contractor and Sub-Contractor, whether arising during the execution or after the completion or abandonment of the Sub-Contract Works or after the determination of the employment of the Sub-Contractor under Sub-Contract NSC/4a (whether by breach or in any other manner), in regard to any matter or thing of whatsoever nature arising out of the Sub-Contract or in connection therewith, then either party shall give to the other notice in writing of such dispute or difference and such dispute or difference shall be and is hereby referred to the arbitration and final decision of a person to be agreed between the parties, or, failing such agreement within 14 days after either party has given to the other a written request to concur in the appointment of an Arbitrator, a person to be appointed on the request of either party by the President or a Vice-President for the time being of the Royal Institution of Chartered Surveyors or, alternatively, if the party requiring the appointment so decides, by the President or a Vice-President for the time being of the Royal Institute of British Architects.

3.2 .1 Provided that if the dispute or difference to be referred to arbitration under this Sub-Contract raises issues which are substantially the same as or connected with issues raised in a related dispute between

the Contractor and the Employer under the Main Contract or

the Sub-Contractor and the Employer under Agreement NSC/2 or NSC/2a as applicable or

the Contractor and any other nominated sub-contractor under Sub-Contract NSC/4 or NSC/4a as applicable or

the Contractor and any Nominated Supplier whose contract of sale with the Contractor provides for the matters referred to in clause 36.4.8 of the Main Contract Conditions

and if the related dispute has already been referred for determination to an Arbitrator, the Contractor and the Sub-Contractor hereby agree that the dispute or difference under this Sub-Contract shall be referred to the Arbitrator appointed to determine the related dispute; and such Arbitrator shall have power to make such directions and all necessary awards in the same way as if the procedure of the High Court as to joining one or more defendants or joining co-defendants or third parties was available to the parties and to him.

.2 Save that the Contractor or the Sub-Contractor may require the dispute or difference under this Sub-Contract to be referred to a different Arbitrator (to be appointed under this Sub-Contract) if either of them reasonably considers that the Arbitrator appointed to determine the related dispute is not appropriately qualified to determine the dispute or difference under this Sub-Contract.

.3 Articles 3.2.1 and 3.2.2 shall apply unless in the Appendix, part 2 Section B the words "Articles 5.1.4 and 5.1.5 apply" have been deleted.

3.3 Such Arbitrator shall not without the written consent of the Contractor and Sub-Contractor enter on the arbitration until after the practical completion or abandonment of the Main Contract Works, except to arbitrate:

3.3 .1 whether a payment has been improperly withheld or is not in accordance with the Sub-Contract; or

3.3 .2 whether practical completion of the Sub-Contract Works shall be deemed to have taken place under clause 14.2; or

3.3 .3 in respect of a claim by the Contractor or counterclaim by the Sub-Contractor to which the provisions of clause 24 apply in which case the Arbitrator shall exercise the powers given to him in clause 24; or

3.3 .4 any matters in dispute under clause 4.3 in regard to reasonable objection by the Sub-Contractor or under clauses 11.2 and 11.3 as to extension of time.

3.4 In any such arbitration as is provided for in article 3 any decision of the Architect which is final and binding on the Contractor under the Main Contract shall also be and be deemed to be final and binding between and upon the Contractor and the Sub-Contractor.

3·5   Subject to the provisions of clauses 4·6, 35·4·3, 36·5·3, 37·5 and clause 30 of the Main Contract Conditions the Arbitrator shall, without prejudice to the generality of his powers, have power to direct such measurements and/or valuations as may in his opinion be desirable in order to determine the rights of the parties and to ascertain and award any sum which ought to have been the subject of or included in any certificate and to open up, review and revise any certificate, opinion, decision, requirement or notice and to determine all matters in dispute which shall be submitted to him in the same manner as if no such certificate, opinion, decision, requirement or notice had been given.

3·6   The award of such Arbitrator shall be final and binding on the parties.

3·7   Whatever the nationality, residence or domicile of the Employer, the Contractor, the Sub-Contractor, and any sub-contractor or supplier or the Arbitrator, and wherever the Main Contract Works or Sub-Contract Works, or any parts thereof, are situated, the law of England shall be the proper law applicable to the Sub-Contract and in particular (but not so as to derogate from the generality of the foregoing) the provisions of the Arbitration Acts, 1950 (notwithstanding anything in S.34 thereof) to 1979 shall apply to any arbitration under article 3 wherever the same, or any part of it, shall be conducted. **[b]**

Attestation   (A)   Signed by or on behalf of the Contractor in the presence of:

CONTRACTOR - STAGE 4

    Signed by or on behalf of the Sub-Contractor in the presence of: **[c]**

SUB-CONTRACTOR - STAGE 4

(B)   The Common Seal of the Contractor was hereunto affixed in the presence of:

(C.S.)

    The Common Seal of the Sub-Contractor was hereunto affixed in the presence of: **[d]**

(C.S.)

Footnotes

[b] Where the parties do not wish the proper law of the Sub-Contract to be the law of England and/or do not wish the provisions of the Arbitration Acts 1950 to 1979 to apply to any arbitration under the Sub-Contract held under the procedural law of Scotland (or other country) appropriate amendments to article 3·5 should be made.

[c] Footnote not applicable.

[d] Footnote not applicable.

Appendix to NSC/4a

part 1

First recital

Particulars of the Sub-Contract Works

Numbered Documents annexed to Sub-Contract NSC/4a
(to be listed here)

First recital

part 2 Section A

Main Contract Works:[n]

CONTRACTOR

Form of Main Contract Conditions:	Standard Form of Building Contract, 1980 edition.
	*Local Authorities/Private edition/WITH/WITH APPROXIMATE/ WITHOUT Quantities (revised . . . .)
	*Sectional Completion Supplement

Inspection of Main Contract:	The unpriced *Bills of Quantities/Bills of Approximate Quantities/ Specification (which incorporate the general conditions and preliminaries of the Main Contract) and the Contract Drawings may be inspected at:

Execution of Main Contract:	*under hand/under seal

Main Contract Conditions – alternative provisions:	Architect/Supervising Office: *Article 3A/Article 3B
	WITHOUT quantities edition only, Quantity Surveyor: *Article 4A/Article 4B
	Master programme: clause 5·3·1·2 *deleted/not deleted
	Works insurance: *clause 22A 22B 22C
	Insurance: clause 21·2·1 Provisional sum *included/not included

Main Contract Conditions – any changes from printed Standard Form identified above:

Footnote

[n] Insert the same description as in the Main Contract Articles of Agreement

*Complete as applicable

part 2 Section B

Main Contract: Appendix and entries therein

	Clause etc	
Statutory tax deduction scheme –Finance (No.2) Act 1975	Fourth recital and 31	Employer at Date of Tender *is a 'contractor'/is not a 'contractor' for the purposes of the Act and the Regulations.
Settlement of disputes– Arbitration	Article 5·1	Articles 5·1·4 and 5·1·5 apply (see Article 5·1·6)
Date for Completion [o]	1·3	_____
Defects Liability Period (if none other stated is 6 months from the day named in the Certificate of Practical Completion of the Works).	17·2	_____
Insurance cover for any one occurrence or series of occurrences arising out of one event	21·1·1	£ _____
Percentage to cover professional fees	22A	_____
Date of possession	23·1	_____
Liquidated and Ascertained Damages	24·2	at the rate of £ _____ per _____
Period of Delay: by reason of loss or damage caused by any one of the Clause 22 Perils	28·1·3 28·1·3·2	_____
for any other reason	28·1·3·1 and 28·1·3·3 to ·3·7	_____
Period of Interim Certificates (if none stated is one month)	30·1·3	_____
Retention Percentage (if less than five per cent)	30·4·1·1	_____
Period of Final Measurement and Valuation (if none stated is 6 months from the day named in the Certificate of the Works)	30·6·1·2	of Practical Completion
Period for issue of Final Certificate (if none stated is 3 months)[p]	30·8	_____
Work reserved for Nominated Sub-Contractors for which the Contractor desires to tender	35·2	_____ _____ _____
†Fluctuations	37	*clause 38/clause 39/clause 40
Percentage addition	*38·7/39·8	_____

CONTRACTOR

Footnotes

[o] If a later Completion Date has been fixed under clause 25 this should also be stated here.

[p] Standard Form Local Authorities Edition only.

†Clause 40 and Formula Rules entries are not applicable where in part 2, Section A it is stated that the WITHOUT Quantities Conditions apply.

*Delete as applicable

NSC/4a

Main Contract: Appendix *continued*

†Formula Rules   40·1·1·1

                   rule 3   Base Month _____19____

                   rule 3   Non-Adjustable Element[q] _____%
                                     (not to exceed 10%)

                   rule 10   Part I/Part II of Section 2 of the Formula Rules is to apply.

part 2 Section C
Obligations or restrictions imposed by the Employer not covered by Main Contract Conditions
(e.g. in Preliminaries in the Contract Bills)[r]

Order of Works: Employer's requirements affecting the order of the Main Contract Works (if any)

Location and type of access:

CONTRACTOR

part 3 _____

Clause 7·1   Insurance cover for any one occurrence or series of occurrences arising out of
one event £ _____

part 4* _____

Clause 11·1   The period/periods when Sub-Contract Works can be carried out on site:

to be between _____   and _____

Period required by Architect to approve drawings   _____
after submission

Periods required:

(1) for submission of all further
     sub-contractor's drawings etc. (*co-
     ordination, installation, shop or builder's
     work or other as appropriate*)[s]   _____

(2) for execution of Sub-Contract Works: off-site   _____

                                 on-site   _____

(3) Notice required to commence work on site   _____weeks

*to be completed as far as relevant

[q] Standard Form Local Authorities WITH Quantities Edition only.

[r] This information, unless included in the Sub-Contract Specification or Bills of Quantities, should be given e.g. by repeating here or by attaching a copy of the relevant section of the Preliminaries Bill of the Main Contract Bills (or of the Main Contract Specification).

[s] Not including period required by Architect for approval.

†See footnote previous page.

part 5

Clause 16·3·4
Clause 17·4·3

The daywork percentages are:

Definition	Labour %	Materials %	Plant %
RICS/NFBTE			
RICS/ECA			
RICS/ECA (Scotland)			
RICS/HVCA			

Clause 21·5

Retention Percentage _____ %

To be the same as that set out or referred to in part 2 Section B – Main Contract, Appendix and entries therein (Main Contract Conditions reference clause 30·4·1·1).

Clause 24

part 6

24·1·2   The Adjudicator is:

24·3·1·2   The Trustee-Stakeholder is:

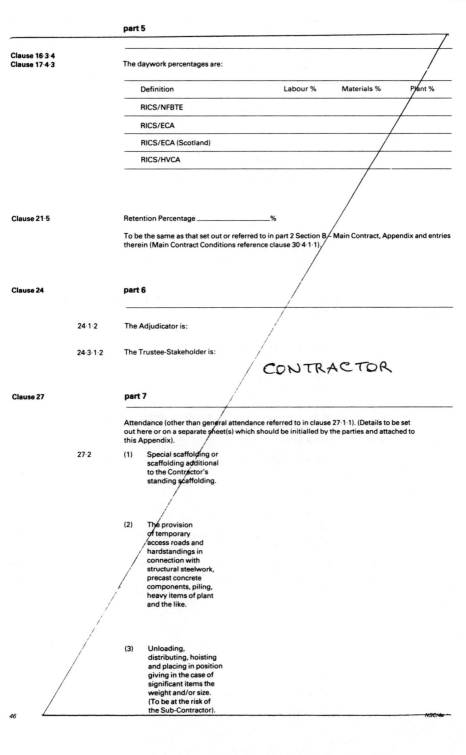

CONTRACTOR

Clause 27

part 7

Attendance (other than general attendance referred to in clause 27·1·1). (Details to be set out here or on a separate sheet(s) which should be initialled by the parties and attached to this Appendix).

27·2   (1)   Special scaffolding or scaffolding additional to the Contractor's standing scaffolding.

(2)   The provision of temporary access roads and hardstandings in connection with structural steelwork, precast concrete components, piling, heavy items of plant and the like.

(3)   Unloading, distributing, hoisting and placing in position giving in the case of significant items the weight and/or size. (To be at the risk of the Sub-Contractor).

NSC/4a

180

(4) The provision of
covered storage and
accommodation
including lighting
and power thereto.

(5) Power supplies giving
the maximum load.

(6) Maintenance of
specific temperature
or humidity levels.

(7) Any other attendance
not included under
(1) to (6) or as general
attendance under
clause 27·1·1.

CONTRACTOR

Clause 34 **part 8**

Fluctuations: *clause 35/clause 36/clause 37

Clause 35 **part 9**

35·2·1 List of materials, goods, electricity and fuels. [t]
(Details to be set out on a separate sheet(s) which should be initialled by the parties and attached
to this Appendix)

35·6·1 Date of Tender

35·7 Percentage _____ %

Clause 36 **part 10**

36·3·1 Materials, goods, electricity and fuels.[t]
–List of basic prices
(Details to be set out on a separate sheet(s) which should be initialled by the parties and attached
to this Appendix)

36·7 Date of Tender

36·8 Percentage _____ %

Footnote [t] Where the Contractor is instructed by the Architect to include
fuels.

*Delete as applicable

NSBI4c 47

Clause 37   **part 11**

(1)   37·1   Nominated Sub-Contract Formula Rules are those dated _____ 19_____
                 *Part I/Part III of these Rules applies

(2)   37·3·3   Non-Adjustable Element[u] _____ % (not to exceed 10%)
      and
      ·3·4

(3)   37·4   List of Market Prices

(4)   JCT Nominated Sub-Contract Formula Rules

rule 3   Definition of Balance of Adjustable Work —
                any measured work not allocated to a Work Category

rule 3   Base Month _____

rule 3   Date of Tender _____

rule 8   Method of dealing with 'Fix-only' work

rule 11(a)   Part I only: the Work Categories applicable to the Sub-Contract Works

rule 43   Part III only: Weightings of labour and materials – Electrical Installations or Heating,
                Ventilating and Air Conditioning Installations

	Labour	Materials
Electrical	_____ %	_____ %
Heating, Ventilating and Air Conditioning[v]	_____ %	_____ %
	_____ %	_____ %

CONTRACTOR

rule 61a   Adjustment shall be effected
                 *upon completion of manufacture of all fabricated components
                 *upon delivery to site of all fabricated components

rule 64   Part III only: Structural Steelwork Installations:
                (i) Average price per tonne of steel delivered to fabricator's work
                £ _____

                (ii) Average price per tonne for erection of steelwork
                £ _____

rule 70a   Catering Equipment Installations:
                apportionment of the value of teach item between
                (i) materials and shop fabrication   £ _____

                (ii) supply of factor items   £ _____

                (iii) site installations   £ _____

Footnotes

[u] Only applicable where the Main Contract is let on the Standard Form of Building Contract, Local Authorities Edition, WITH Quantities.

[v] If both specialist engineering formulae apply to the Sub-Contract the percentages for use with each formula should be inserted and clearly identified. The weightings for sprinkler installations may be inserted where different weightings are required.

182

part 12 (See Clause 20A, footnote[i])

Clauses 20A and 20B

Finance (No.2) Act 1975 – Statutory Tax Deduction Scheme

1. The Contractor* is/is not entitled to be paid by the Employer without the statutory deduction referred to
in the above Act or such other deduction as may be in force:

2. The evidence to be produced to the Contractor for the verification of the Sub-Contractor's tax certificate (expiry date   19   ) will be:

CONTRACTOR

part 13

Any other matters (e.g. special conditions or agreements on employment of labour, limitation on working hours) to be set out here after agreement between the Contractor and Sub-Contractor on these matters:

*Delete as applicable

Explanatory notes

The Nominated Sub-contract NSC/4a differs from NSC/4 in the respect that the entire Sub-contract is in one document, consisting of the Articles of Sub-contract Agreement, the Conditions and the Appendix. Here an appendix is necessary as there is no tender form in the Alternative Method. As the Sub-contract Conditions are the subject of a companion book in this series and discussed in detail therein, only the Articles of Sub-contract Agreement and the Appendix are discussed here.

Articles of Sub-contract Agreement

After the date and the names and addresses of the parties to the agreement have been completed, there are five recitals.

First   Here it is stated that the Sub-contractor has submitted an offer and further states that all documents forming that offer are to be numbered for identification in the Appendix. The difference here from the Basic Method is that the Proposed Sub-contractor does not have the benefit of Tender NSC/1 where all the requirements of the Sub-contract are set out. Therefore, the importance of the Sub-contractor having full knowledge of the Main Contract Conditions and how the Appendix to NSC/4a will be completed, cannot be overlooked. A full description should of course have been given to the tenderer at the time of his making an offer.

Second   Here the Employer's name is entered and the Main Contract Works are stated to be described in the Appendix.

Third   Whether or not completion of Agreement NSC/2a has been made is stated here and deleted if the Agreement has not been completed.

Fourth   This recital is important as it forms the start date for the 14-day period by which this contract has to be executed. It also confirms that a nomination has taken place, stating the date of that nomination.

Fifth   By deletion the tax status of both the Sub-contractor and the Contractor is confirmed, together with the status of the Employer as regards his position as a contractor.

Articles

1   The Sub-Contractor accepts his obligations to carry out and complete the Sub-contract works in accordance with NSC/4a and his offer, all as described in the Numbered Documents.
2   By deletion, it is established whether the offer is a Sub-contract Sum or a Tender Sum and that it includes a 2½ per cent cash discount to the Contractor. Great care must be taken here to enter both the correct sum and the exact amount.
3   This is exactly the same as NSC/4 and again reflects the arbitration arrangements as set out for the Main Contract.

The attestation has to be signed in the presence of two witnesses to each signature and can be either under hand or under seal.

Appendix

Part 1   The particulars of the Sub-contract Works described here will contain any reference to design work, performance specification or selection of materials, together with the description of the trade or specialism.

Probably the space for the list of Numbered Documents will be insufficient and a separate page may have to be attached, suitably titled and signed by both parties. These numbered documents are important; each document has a separate number and they, together with this Sub-contract NSC/4a, are the Contract Documents.

Part 2   The Main Contract Works are the same description as in the Main
Section A   Articles of Agreement.

The Form of Main Contract Conditions are identified by deleting the inappropriate forms, and again this should accord with the Main Articles of Agreement.

Ideally all documents for inspection should be at the one address, but the opportunity is given here to state where they can be inspected.

It is stated whether the Main Contract is executed under hand or under seal, by deletion.

The alternative provisions which may be applicable in the Main Contract should be mirrored here exactly.

Any modifications or amendments to the Standard Form should be entered here, such as a performance bond or any other additional conditions.

Note that all the above information will have been given to the Sub-contractor at tender stage, and it governs the Sub-contract, whether accurately copied or not. (Clause 2.2 NSC/4a).

Section B   This should be a careful copy of the Main Contract Appendix.

Section C   Here a copy of the unpriced Preliminaries Bill would be the easiest way to transmit this information, with the items in question referred to and marked for reference. The Employer may impose restrictions or obligations upon the Contractor as to ingress and egress, use of the site, or the order in which the works are to be carried out. If the Architect issues instructions which vary this information detailed here, they will become variations and the Sub-contractor could make reasonable objection to variations of these provisions.

Part 3   As there is no Tender NSC/1 this has to be agreed between the Contractor and the Sub-contractor when NSC/4a is being executed. Reference is made to Clause 7.1 where the Sub-contractor has to insure for his liability in respect of personal injury and death and for his liability in respect of injury and damage to property, as detailed in the Appendix.

There will also have to be agreement on the minimum insurance cover for any one occurrence or series of occurrences arising out of one event.

Part 4   In the reading of Clause 11.1 the Sub-contractor shall carry out and complete the works in accordance with the agreed programme details in the Appendix – Part 4, subject to receipt of the notice to commence work on site as detailed in the Appendix – Part 4.

Notice that the wording is 'agreed programme details'; thus the periods here have to be agreed between the Contractor and the Sub-contractor before execution of NSC/4a. If the Sub-contract has no design element and no drawings are required, some entries will be inapplicable.

Note that the period required for submission of all further drawings does not include the period for the Architect's approval. See the footnote.

Part 5   The Sub-contractor has to enter his daywork percentages here, as required by Clauses 16.3.4.1 and 17.4.3.1 of NSC/4a.

The retention percentage will be the same as the Main Contract.

Part 6   Again these names will be by agreement of the parties.

Part 7   These are the agreed special attendances required by the Sub-contractor, and reference should be made to the explanatory notes for Tender NSC/1 as to the meanings of the various special attendances.

Part 8   The Sub-contractor will have been informed at tender stage which clause is to apply for fluctuations. Two clauses have to be deleted and if this is not done then Clause 35 is deemed to operate.

Part 9   At tender stage the Sub-contractor will have received from the Architect the date of tender, percentage addition and also the information as to whether or not 'fuels' is allowable.

Also at tender stage the Sub-contractor will have submitted a list of the materials, etc., and fuels if allowed, to which he wishes fluctuations to apply. This list is restrictive and Clause 35 will apply only to those materials, etc. as listed.

Part 10   This is the alternative Clause 36 where it applies, and again the Sub-contractor will have been informed at tender stage of the date of tender, percentage addition, and also whether or not 'fuels' is allowed.

Also at tender stage the Sub-contractor will have submitted a list of basic prices to which he wishes Clause 36 to apply, and fuels if allowed.

Note that here, unlike Tender NSC/1, there is no form set out for the list and it is suggested that separate sheets be attached, each initialled by the parties. The four-column list as set out in Tender NSC/1 would be an ideal format to follow, completed as necessary with the materials listed, both by name and their basic prices current at date of tender. Again, this list is restrictive and only those materials named with their basic prices qualify for Clause 36 fluctuations.

Part 11   This is the alternative Clause 37 where the 'formula method' operates for fluctuations. At tender stage the Architect will have informed the Sub-contractor of the entries in this page, with the exception of Clause 37.4, Rule 64, Rule 70a and Clauses 20A and 20B, all of which will be entered by the Sub-contractor.

Clause 37.1   The current edition of the Nominated Sub-contract Formula Rules are entered here, together with Part I or Part III, whichever is applicable. Part I is the Work Category Method and Part III is for specialist engineering work being carried out by the Sub-contractor.

Clauses 37.3.3 This applies only to the Local Authorities Edition with Quantities
and 37.3.4   where the percentage is deducted from the amount of the fluctuations. The reasoning being that a Contractor should not recover fluctuations on his overhead expenses.

Clause 37.4   As in the Main Contract, this is a list of market prices current at date of tender which refers to goods and materials manufactured outside the United Kingdom, imported and to be fixed into the works without further processing. This list is excluded from formula adjustment and can be varied only by a direct change in market prices. Any tax considerations, excepting VAT, are also included.

Rule 3   The sub-contractor will have been informed at tender stage of the items which cannot be allocated properly to any one of the work categories and are listed here before the execution of NSC/4a.
                  The Base Month is normally the calendar month prior to that in which the Tender is to be returned. In this case the Architect will be careful to insert the Base Month with the tender submission date in mind. There is an option to change this method of calculation if the Tenderer was informed otherwise.
                  The date of tender is that ten days before the date for return of tenders. Again the Architect will be careful that he reflects the correct date in relation to competitive tenders, more especially if a preliminary tender was obtained. This would be the date at which the Tender Sum was prepared.

Rule 8   At tender stage the Sub-contractor will have been informed as to how fix-only work will be dealt with, in one of three options, as follows:

                  Including it in an appropriate work category
            OR   Including it in the Balance of Adjustable Work
            OR   By the use of the fix-only index

Rule 11a   When only Part 1 applies, a list is to be attached to the Tender in order to classify each item in the Sub-contractor's Bill of Quantities or Schedule of Rates. The classifications are

                  Work category
                  Fix-only work
                  Specialist engineering work

Provisional sum work
Balance of adjustable work
Work excluded from formula adjustment

Rule 43   When only Part III applies, the labour and material percentages have to be entered and should, of course, total 100 per cent in each line.

Rule 61a   This applies to shop fabrication, and one alternative will be deleted at tender stage by the Architect.

Rule 64   These average prices are submitted by the Sub-contractor with his tender.

Rule 70a   Again the Sub-contractor will have submitted these apportionments with his Tender.

Part 12   This refers to the Sub-contractor's status in respect to the Finance
Clauses 20A   (No 2) Act 1975 – Statutory Tax Deduction Scheme. In (1) the
and *20B*   entitlement to deduction or not of tax, is stated, which means simply, 'does he hold a current tax ememption certificate?' In (2), if a certificate is held, the expiry date is to be stated and the Sub-contractor here will state how he means to verify his status. In accordance with Clause 20A of NSC/4a, the production of a current tax certificate not later than 21 days before the first payment is due, and in turn the Contractor's confirmation within a further seven days, satisfies the Conditions.

Part 13   After agreement between the Contractor and the Sub-contractor, any matters falling under this heading are listed here. These matters might be, if the Sub-contractor wishes to carry out work during local holidays or at other public holiday times, when the Contractor is not on site. Or anything else which could affect working hours or the employment of labour.

Payment of Nominated Sub-contractor (Main Form)

35.13.1   When a value is included in an Interim Certificate for any nominated sub-contractor, it shall be calculated in accordance with Clause 21.4 of Sub-contract NSC/4 or NSC/4a. The Architect must inform the Contractor and each applicable Nominated Sub-contractor of the amount included in the Certificate.

35.13.2   The Contractor must pay the Nominated Sub-contractor their amounts as included in the Interim Certificates within 17 days of the date of issue of the certificate by the Architect. The amount included in the certificate might not be the same as that paid by the Contractor, due to deductions of cash discounts or other deductions empowered by the Conditions, i.e. 'set-off'.

35.13.3   The Nominated Sub-contractor must give to the Contractor some form of receipt which he in turn can pass on to the Architect as proof of payment.

35.13.4   If the Sub-contractor does not provide the Contractor with proof of payment the latter cannot then provide proof to the Architect. If the Architect is satisfied that this is the only reason why the Contractor does not provide proof of payment, he will take the matter no further and deem that for the purpose of this clause proof has been provided.

35.13.5   The Employer has the option to operate the two preceding Clauses 35.13.5.3 and 35.13.5.4, but if Agreement NSC/2 or NSC/2a has been executed, it is mandatory for these clauses to be operated.

If reasonable proof of payment is not provided to the Architect before the issue of an Interim Certificate or the Final Certificate, and failure of the Nominated Sub-contractor to produce receipts is not substantiated as the reason, the Architect issues a certificate. The certificate states that the Contractor has failed to produce reasonable proof of payment, names the Sub-contractor involved, gives the amount of payment and issues it direct to the Employer with a copy to the Nominated Sub-contractor.

Note that for every affected sub-contractor, a separate certificate is necessary.

Always provided that the Architect has issued the required certificate or certificates, then

future payments due to the Contractor, less any amounts due to the Employer, will be reduced by

any amounts due to Nominated Sub-contractors which the Contractor has not paid, together with any VAT involved. Here the Nominated Sub-contractor will have to help and account for any amounts of VAT which those payments would attract.

The Employer will pay the Nominated Sub-contractor direct, provided that there is sufficient monies due to the Contractor to cover the direct payment.

A direct payment to a nominated sub-contractor is subject to the following:

(1)   The Nominated Sub-contractor will be paid by the Employer at the same time as the reduced amount is paid to the Contractor. If there is no balance left to be paid to the Contractor, the Nominated Sub-contractor will still have his amount paid within the normal 14 days applicable to any Interim Certificate.

(2)   Where the sum due to the Contractor is all retention or part retention, the reduction and payment to the Nominated Sub-contractor shall not exceed any part of the Contractor's retention which would otherwise have been due for payment.

(3)   Where the Employer has to make payments direct to two or more nominated sub-contractors, and the amount due to the Contractor is not sufficient to pay all the sub-contractors in full, he will *pro rata* the amount available to the amounts from time to time unpaid by the Contractor, or adopt some other fair and reasonable means of apportionment.

(4) The need to make direct payments, whether mandatory or discretionary, ceases if there is a petition to the Court for the winding-up of the Contractor, or a resolution passed for the winding-up of the Contractor (other than for the purposes of amalgamation or reconstruction) whichever occurs first. Note that where the Contractor is subject to bankruptcy, then the clause will need amendment to refer to the events of the happening of which bankruptcy occurs.

As the Employer can make direct payments at his discretion or it is mandatory for him to do so, dependent upon whether or not Agreement NSC/2 or NSC/2a has been executed, it might be argued that the wording 'fair and reasonable having regard to all the circumstances' might create two classes of nominated sub-contractor, as regards their right to direct payment. The discretionary payment being second to the preferential payees of the mandatory arrangement, when payments are apportioned.

Clause 7.2 of Agreement NSC/2 and Clause 6.2 of Agreement NSC/2a provide that the Nominated Sub-contractor will repay to the Employer any direct payment, if the Employer produces reasonable proof that there was in existence at the time of such payment, a petition or resolution in terms of Clause 35.13.5.4.4 of the Main Contract Conditions.

Where neither NSC/2 nor NSC/2a has been executed, both the Employer and the Nominated Sub-contractor might well enter into an arrangement to have the same terms as above operating, conditional upon discretionary payments being made by the Employer. This would mean that the Employer, in these circumstances, would also recover his payments in the case of the winding-up of the Contractor.

Extension of period or periods for completion of Nominated Sub-contract Works

35.14   No extension of time will be granted to the Nominated Sub-contractor by the Contractor without the Architect's written permission, in accordance with NSC/4 or NSC/4a and as referred to in Clause 11.2.2 of these documents. The programme for Nominated Sub-contract Works and completion date is to be found in Schedule 2, Item 1c of Tender NSC/1.

Under Clause 11.2.2 of Sub-contract NSC/4 or NSC/4a, the Architect has the duty to consent in writing to a grant by the Contractor of an extension of time to the Nominated Sub-contractor. Notices are received from the Nominated Sub-contractor through the Main Contractor that the delay was the fault of the Main Contractor or the occurrence of a Relevant Event. Although the delay was the fault of the Main Contractor, he must pass all notices on to the Architect.

Failure to complete Nominated Sub-contract Works

35.15   The Contractor will notify the Architect (with a copy to the Sub-contractor) if the Sub-contractor fails to complete his works within the agreed period or extended period of the Sub-contract. The Architect, if satisfied, will certify to the Contractor the failure to complete (with a copy certificate to the Sub-contractor). The issue of this Certificate allows the Contractor to recover damages from the Sub-contractor for loss and expense caused by his failure to complete (Clause 12.2 of NSC/4 or NSC/4a).

As shown in Chapter 4, this is the only certificate which is issued directly to the Contractor (Clause 5.8).

The Architect has a period of two months in which to issue the certificate, starting from the date of the Contractor's original notification.

Practical Completion of Nominated Sub-contract Works

35.16   When the Architect is satisfied that Practical Completion of any nominated sub-contract work is achieved, he issues to the Employer, a certificate to that effect, with a copy to the Contractor and the Nominated Sub-contractor. A separate certificate is required for every nominated sub-contract works.

This certificate means to the Sub-contractor that the Defects Liability Period has started, the sequence of early final payments by the Employer starts, and the making good of defects as detailed in Clause 14.3 of NSC/4 comes into operation.

Final payment of Nominated Sub-contractors

35.17   Where Agreement NSC/2 or NSC/2a has been entered into and Clause 8 or Clause 7 of each Agreement respectively remain unaltered, the Architect may issue an Interim Certificate which includes the final balance due to a Nominated Sub-contractor. However, 12 months after the date of the Certificate of Practical Completion, the Architect must issue such an Interim Certificate. This is the early final payment, but the following provisions must be observed:

(a)   All defects, shrinkages or other faults, which are the responsibility of the Sub-contractor to remedy, have been made good to the Architect's satisfaction.

(b)   All documents necessary for ascertaining the final sum of the Sub-contract have been sent through the Main Contractor to the Architect or Quantity Surveyor.

(c)   The Certificate of Practical Completion has been issued.

Clause 8 and Clause 7 of Agreements NSC/2 and NSC/2a respectively state that the Architect must operate the provisions of Clause 35.17; thus, if the Architect defaults in certifying the early final payment, then the Sub-contractor has a direct legal right against the Employer. This right only comes into being 12 months after the date of the issue of the Certificate of Practical Completion when the Architect is obliged to certify payment.

The same clause puts an obligation upon the Sub-contractor to rectify any faults after his final payment and before the Main Contract Final Certificate is issued and for the period of the Main Contract Defects Liability Period.

35.18   This clause covers for the failure of the Nominated Sub-contractor to make good any defects, shrinkages, etc., before the issue of the Main Contract Final Certificate and subsequent to his early final payment. If this happens, the Architect must nominate another or substitute nominated sub-contractor to remedy the defects (Clause 35.24 refers).

The Contractor is given the opportunity to examine and agree to the Sub-contract price of the second Nominated Sub-contractor and this agreement will not be unreasonably withheld.

The Employer will take all reasonable steps, through Agreement NSC/2 or NSC/2a, to recover the additional costs of employing a second Nominated Sub-contractor. Any amounts not recovered have to be borne by the Main Contractor, if he has had the opportunity to examine and agree the second Nominated Sub-contractor's Sub-contract Sum.

Nothing in Clause 35.18 shall override or modify the provisions of Clause 35.21.

Clause 35.21 clearly states that the Contractor's responsibility covers defects in materials and/or workmanship only and he has no responsibility for any defects because of faults, errors or omissions in the design element by the Nominated Sub-contractor. These latter defects are the direct responsibility of the Nominated Sub-contractor to the Employer.

35.19   Regardless of whether or not a final payment has been made to a Nominated Sub-contractor, until the date of Practical Completion or possession by the Employer, whichever is first, the Contractor is still responsible for loss or damage to the Works or materials and goods.

Whichever insurance clause is applicable, 22A, 22B or 22C, remains in force.

Position of Employer in relation to Nominated Sub-contractor

35.20   The Employer's liability and legal relationship to the Nominated Sub-contractor is only by virtue of Agreement NSC/2 or NSC/2a. No other legal relationship operates because of Clause 35.

Clause 2 of Agreement NSC/2 or Clause 1 of Agreement NSC/2a – Position of Contractor (Main Form)

35.21   As already stated, clearly the Contractor has no responsibility with regard to the design element or selection of materials by a nominated sub-contractor, whether Agreement NSC/2 or NSC/2a has been executed or not. Therefore, if no agreement has been executed, the Employer has no redress or protection against the Nominated Sub-contractor's errors or short-comings in design and material selection.

If a design element is required in a nominated sub-contract, this protection and legal right of redress is important to the Employer. To obtain it, the nomination must be by the Basic Method, or if by the Alternative Method then on the basis that Agreement NSC/2a is to be executed. This will obviously be a very important consideration on the Architect's part when deciding upon nomination.

Restrictions in Contracts of Sale, etc – Limitation of liability of Nominated Sub-contractors

35.22   Clause 2.3 of Sub-contract NSC/4 or NSC/4a clearly states that if the Sub-contractor obtains goods or services from a supplier or sub-sub-contractor, and limitations, restrictions or exclusions apply to these goods or services, the Sub-contractor will inform the Contractor who, in his turn, informs the Architect.

If the Contractor and Architect approve these limitations, restrictions or exclusions, then the same will apply to the Sub-contractor's liability to the Contractor.

Unless the Contractor and Architect approve, the Sub-contractor does not need to enter into any such sub-sub-contract.

Position where proposed nomination does not proceed further

35.23   The Architect has the choice either to issue an instruction requiring a variation for the omission of the work for which he intended to nominate a sub-contractor, or to select another person to be nominated under Clause 35, if

(1)   The Contractor objects to proposed sub-contractor (*see Diagram 6*, page 124, and *Diagram 9*, page 162).

or (2)   Proposed sub-contractor withdraws because he is unable to agree the particular Conditions in Schedule 2 of Tender NSC/1, Basic Method (*see Diagram 7*, page 125).

or (3)   The proposed sub-contractor fails to enter into Sub-contract NSC/4a, Alternative Method (*see Diagram 9*, page 162).

or (4)   The proposed sub-contractor withdraws (*see Diagram 5*, page 123).

Circumstances where re-nomination is necessary (Clause 35.24 Main Form)

The following Table shows the causes of re-nomination and the action to be taken:

Cause	Action
1 Wholly suspends the carrying out of the works (Clause 29.1.1 NSC/4 or 4a) 2 Fails to proceed with works as agreed (Clause 29.1.2 NSC/4 or 4a) 3 Refuses to remove defective work or rectify defects (Clause 29.1.3 NSC/4 or 4a) 4 Assigns or sub-lets the Contract without permission (Clause 29.1.4 NSC/4 or 4a) 5 Fails to comply with Fair Wages Clause, if applicable (Clause 29.1.4 NSC/4 or 4a)	Architect issues instruction to Contractor to give notice to Sub-contractor of his default and may also tell Contractor to wait for further instruction before determining Sub-contractor's employment (Clause 35.24.4.1) Contractor informs Architect that notice has been given, whether or not he has determined Sub-contractor's employment, or after a further Instruction confirms determination (Clause 35.24.4.2) If Contractor confirms determination by Clause 29.1.3, Architect will renominate and the Contractor given the opportunity to agree the new tender sum (Clause 35.24.4.3)
6 Nominated Sub-contractor becomes bankrupt (Clause 35.24.2 Main Conditions)	Architect re-nominates unless he allows receiver or liquidator to continue the Sub-contract Works (Clause 35.24.5)
7 Nominated Sub-contractor determines his employment because of Contractor's default (Clause 35.24.3 and Clause 30 NSC/4 or 4a)	Architect re-nominates

In cases 1 to 6 inclusive, the Employer is liable for all the additional costs of re-nomination. In case 7 all the additional costs are the Contractor's liability, deductible by the Employer from monies due, or recoverable as a debt.

Determination or determination of employment of Nominated Sub-contractor – Architect's Instructions

35.25   The Contractor cannot determine a nominated sub-contract without an instruction from the Architect. Note that this is the Sub-contract itself and not the Sub-contractor's employment.

35.26   When a nominated sub-contractor's employment is determined under Clause 29 of either NSC/4 or NSC/4a, the Architect will direct the Contractor as to amounts included in Interim Certificates, all in accordance with Clause 29.4 of the Sub-contract.

Until after completion of the Sub-contract, by another re-nominated Sub-contractor, the Contractor need make no further payment to the defaulting Sub-contractor.

After completion of the Sub-contract, the original Sub-contractor can apply to the Contractor for payment, which application is passed to the Architect. The Architect or Quantity Surveyor will value the amount of direct loss and/or expense caused to the Employer by the re-nomination. The Architect will certify the value of work done and of goods and materials supplied and still outstanding.

On payment of the Certificate, the Employer can deduct the amount of his loss and/or expense. The Contractor in turn, when making payment, can deduct a cash discount of 2½ per cent and deduct the amount of his own direct loss and/or expense caused to him by the re-nomination.

Scotland

Clause 35 – Nominated Sub-contractors

For this clause to operate in Scotland a separate set of documents is necessary. The list is included in that of the JCT published documents at the beginning of this book. All the documents have the same numbers as the English/Welsh but the suffix 'Scot' is appended to the numbers to distinguish them. There is one extra document on the list which is not published or used in England and Wales, namely, the Standard Form of Nomination NSC/3a/Scot used at the Architect's discretion when the Alternative Method is employed. There is also an extra clause, 32.27, in the Scottish Supplement, and the layout of the Scottish documents is different.

Nomination procedure – Basic Method

Diagram 1(S) following shows the complete procedure from start to finish, using the Scottish version of the forms.

Diagram 2(S) hows the procedure if the proposed Sub-contractor withdraws.

Diagram 3(S) shows the procedure if the Main Contractor makes reasonable objection.

Diagram 4(S) shows the procedure where the Contractor and the Proposed Sub-contractor are unable to agree within ten working days.

Immediately following the four *Diagrams* are the necessary forms, indicating where the various parties have to provide information and/or signatures to effect a nomination, together with such additional explanatory notes to the version for England and Wales as are necessary.

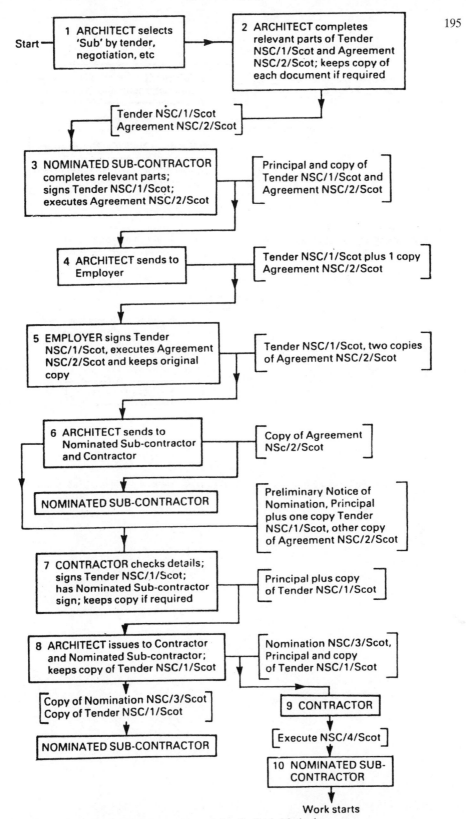

Diagram 1(S): Nomination of a Sub-contractor using the Basic Method

196

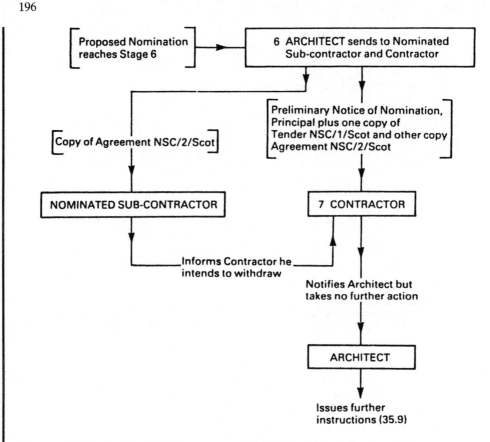

Diagram 2(S): Nomination of a Sub-contractor using the Basic Method – Proposed Sub-contractor withdraws

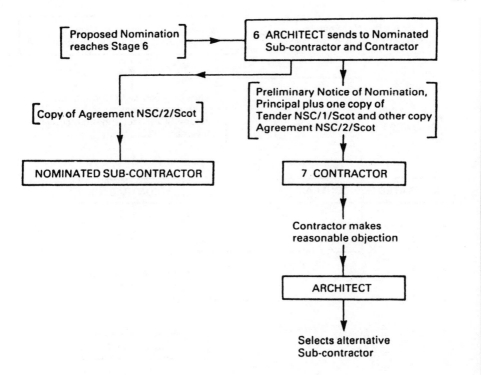

Diagram 3(S): Nomination of a Sub-contractor using the Basic Method – Contractor makes reasonable objection

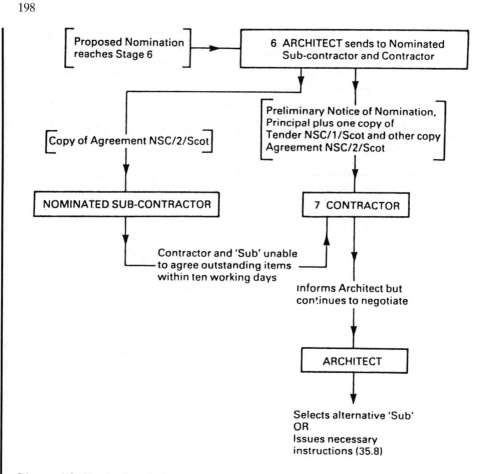

Diagram 4(S): Nomination of a Sub-contractor using the Basic Method – Contractor and Proposed Sub-contractor unable to agree within ten working days

Instructions for Use

Tender NSC/1/Scot

Stage

1. Architect issues a principal and one copy to proposed Nominated Sub-Contractor (NSC) having first completed those parts of the document which require completion by him (i.e. Schedule 1 and relevant part of Appendix (A) if applicable: Item 1A of Schedule 2). Retains photocopies for his own use.

   (Other documents required: Agreement NSC/2/Scot also sent to NSC).

2. NSC completes those parts of the document which require completion by him (i.e. Page 1; Schedule 1 Appendix (A) or (B) and relevant parts of Schedule 2). Signs document on Pages 1 and 5 and 6 or 7 and thereafter returns principal to the Architect;

   (Other documents required: NSC signs Agreement NSC/2/Scot and returns it to Architect).

3. Architect signs Page 1 and then sends principal and one copy to Main Contractor with instruction to agree Schedule 2 with NSC.

   (Other documents required: Employer signs NSC/2/Scot and photocopy sent by Architect to NSC).

4. After reaching agreement with NSC, Main Contractor completes relevant parts of Schedule 2 which is then signed by Main Contractor and NSC on Page 11. Main Contractor also signs document on Page 1 and then returns principal to Architect.

5. Architect makes two photocopies of completed document: Retains one for his own use: Sends the other to NSC and the principal to the Main Contractor.

   (Other documents required: Architect issues Nomination NSC/3/Scot to Main Contractor with copy to NSC: Sends copy of Agreement NSC/2/Scot to Main Contractor).

SBCC

STANDARD FORM

of

NOMINATED SUB-CONTRACT

TENDER

for use in Scotland

_____Employer

_____Sub-Contractor

_____Main Contractor

_____Works

The constituent bodies of the
Scottish Building Contract Committee are:

Royal Incorporation of Architects in Scotland

Scottish Building Employers Federation

Scottish Branch of the Royal Institution of
    Chartered Surveyors

Convention of Scottish Local Authorities

Federation of Specialists and Sub-Contractor
    (Scottish Board)

Committee of Associations of Specialist
    Engineering Contractors (Scottish Branch)

Association of Consulting Engineers
    (Scottish Group)

Copyright of the S.B.C.C., 39 Castle Street, Edinburgh

October 1982

Standard Form of Nominated Sub-Contract Tender for use in Scotland

Main Contract Works:

Location:

Sub-Contract Works: *ARCHITECT - STAGE 2 OR STAGE 1*

To: The Employer and Main Contractor

We,_____

of _____

_____

hereby offer subject to the conditions overleaf to carry out and complete as a Nominated Sub-Contractor the Sub-Contract Works referred to above for

*The VAT exclusive sum of £_____(hereinafter referred to as 'the sub-contract sum')

*Delete as applicable

OR *NOMINATED SUB-CONTRACTOR - STAGE 3*

*The VAT exclusive tender sum† of £_____(hereinafter referred to as 'the sub-contract tender sum')

†For use where the Sub-Contract Works are to be completely remeasured and valued

(Any VAT payable will be dealt with in accordance with the Sub-Contract referred to in (2) below).

in accordance with

(1) The drawings*/specifications*/bills of quantities*/schedule of rates annexed hereto numbered _____ signed by ourselves and the Architect/Supervising Officer.

(2) The Building Sub-Contract known as NSC/4/Scot together with the amendments and modifications contained in the Appendix thereto, to the Standard Conditions of Contract for Nominated Sub-Contractors known as NSC/4,

(3) The particulars of the Main Contract contained in Schedule 1 hereto.

(4) The Particular Conditions set out in Schedule 2 hereto after agreement of the same with the Main Contractor.

(5) The following day work percentages (Sub-Contract NSC/4 Clauses 16.3.4 or 17.4.3) are

Applicable Definition	Labour	Materials	Plant
1. R.I.C.S./N.F.B.T.E._____	_____%	_____%	_____%
2. R.I.C.S./E.C.A./E.C.A. of S _____	_____%	_____%	_____%
3. R.I.C.S./H.V.C.A. _____	_____%	_____%	_____%

We confirm that the sub-contract sum or sub-contract tender sum and the above percentages take into account the 2½% cash discount allowable to the Main Contractor.

NOMINATED SUB-CONTRACTOR - STAGE 3 _____Date

Sub-Contractor

The foregoing offer is approved on the terms and conditions set out above

ARCHITECT - STAGE 5 _____Date

(Approved by the Architect/Supervising Officer on behalf of the Employer)

The foregoing offer is accepted on the terms and conditions set out above

CONTRACTOR - STAGE 7 _____Date

(Accepted by the Main Contractor subject to receipt of a nomination instruction).

Page 1

Tender NSC/1/Scot

Conditions referred to on Page 1

1. We agree to be bound by the Agreement between ourselves and the Employer (NSC/2/Scot) which we have executed simultaneously with this tender after the Architect/Supervising Officer has approved our tender on behalf of the Employer by signature overleaf.

2. If the identity of the Main Contractor is not known to us at the date of our tender we reserve the right within 14 days of written notification by the Architect/Supervising Officer of such identification to withdraw this tender notwithstanding prior approval by or on behalf of the Employer.

3. We reserve the right to withdraw this tender if we are unable to agree the Particular Conditions set out in Schedule 2 hereto with the Main Contractor.

4. Without prejudice to 2 and 3 above this tender shall be deemed to have been withdrawn if the nomination instruction is not issued by the Architect/ Supervising Officer within *_____weeks from the date hereof or such other later date as may be notified by us in writing to the Architect/Supervising Officer.

*Sub-Contractor to insert acceptance period

5. Any withdrawal under 2,3 or 4 above shall be free of cost to the Employer except for any sums that may be due to us in accordance with the Agreement NSC/2/Scot referred to in 1 above.

NOMINATED SUB-CONTRACTOR
OR ARCHITECT - STAGE 3

Page 2

Schedule 1:
Particulars of the Main Contract and Sub-Contract

Employer:

Architect/Supervising Officer:

Quantity Surveyor: ARCHITECT - STAGE 2

Main Contractor:

*Delete as applicable	1.	Sub-Contract Conditions	NSC/4 as modified and amended by NSC/4/Scot all as appropriate to the Standard Form of Building Contract referred to in item 5 hereof (to be executed forthwith after issue of the Architect/Supervising Officer's Nomination NSC/3/Scot).

<table>
<tr><td>***Delete as applicable**</td><td>2.</td><td>Sub-Contract Fluctuations</td><td>NSC/4
*Clause 35 (see Appendix (A)): Clause 35.7____%
*Clause 36 (see Appendix (A)): Clause 36.8____%
*Clause 37 (see Appendix (B)).</td></tr>
<tr><td></td><td>3.</td><td>Main Contract Appendix II & entries therein (see item 10)</td><td>Where relevant these will apply to the Sub-Contract unless otherwise specifically stated here. The entry relating to Clause 37 of the Main Contract is for information of the Sub-Contractor only.</td></tr>
<tr><td></td><td>4.</td><td>Title and address of Main Contract Works</td><td></td></tr>
<tr><td>***Insert the appropriate version**</td><td>5.</td><td>Form of Main Contract Conditions</td><td>Building Contract and Scottish Supplement (1980 Edition) to the Standard Form of Building Contract 1980 Edition*</td></tr>
<tr><td>***Delete as applicable**</td><td>6.</td><td>Inspection of Main Contract</td><td>The unpriced Bills of Quantities*/Bills of Approximate Quantities*/Contract Specification (which incorporate the general conditions and preliminaries of the Main Contract) and the Contract Drawings may be inspected at</td></tr>
<tr><td></td><td>7.</td><td>Execution of Main Contract</td><td>*This item which indicates whether the Main Contract is or is not executed under Seal is not applicable to Scotland but it is retained to maintain the same numerical sequence as in NSC/1/Schedule 1 published by the J.C.T. for use in England.*</td></tr>
<tr><td>***Delete as applicable**</td><td>8.</td><td>Main Contract Conditions alternative provisions</td><td>*Architect/Supervising Officer: Building Contract Clause 3.
Works insurance: *Clause 22A/22B/22C.
Insurance: Clause 21.2.1 Provisional sum included*/not included.</td></tr>
<tr><td></td><td>9.</td><td>Main Contract Conditions – insert here any amendments or additions to Form of Main Contract referred to in item 5</td><td></td></tr>
</table>

Tender NSC/1/Scot **Schedule 1**

Clause etc.

*Delete as applicable

10.	Main Contract: Appendix II and entries therein	Statutory tax deduction scheme – Finance (No. 2) Act 1975	31 Employer at date of tender is a 'contractor'*/is not a 'contractor' for the purposes of the Act and the Regulations
		Defects Liability Period (if none other stated is 6 months from the day named in the Certificate of Practical Completion of the Works)	17.2 _____
		Insurance cover for any one occurrence or series of occurrences arising out of one event	21.1.1 £ _____
		Percentage to cover Professional fees	22A _____
		Date of Possession	23.1 _____
		Date for Completion	1.3 _____
		Liquidate and Ascertained Damages	24.2 at the rate of £ _____ per _____

Period of delay:
by reason of loss or damage caused by any one of the Clause 22 Perils — 28.1.3.2 _____

for any other reason — 28.1.3.1, 28.1.3.3 to .7 _____

Period of Interim Certificates (if none stated is one month) — 30.1.3 _____

Retention Percentage (if less than five per cent) — 30.4.1.1 _____

Period of Final Measurement and Valuation (if none stated is six months from the day named in the Certificate of Practical Completion of the Works) — 30.6.1.2 _____

Period for issue of Final Certificate (if none stated is three months) — 30.8 _____

Work reserved for Nominated Sub-Contractors for which the Contractor desires to tender — 35.2 _____

ARCHITECT - STAGE 2

*Delete as applicable

Fluctuations	37	*Clause 38 *Clause 39 *Clause 40
Percentage addition	38.7 or 39.8	_____ %
Formula Rules Rule 3	40.1.1.1	Base month _____ 19__

*Not to exceed 10%

Rule 3 — *Non-Adjustable Element _____% (Local Authority with quantities only)

*Delete as applicable

Rule 10 and 30(i) — Part I*/Part II of Section 2 of the Formula Rules is to apply.

204

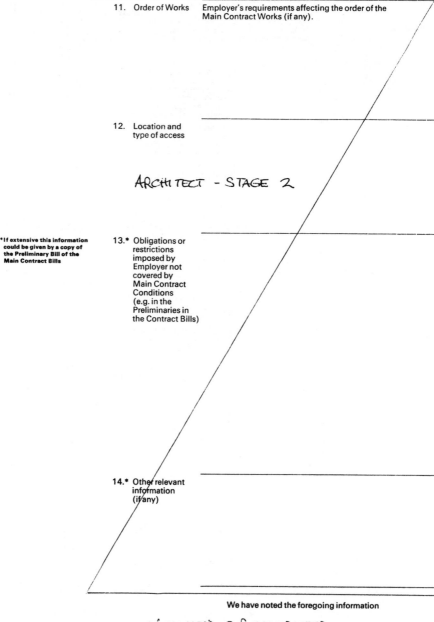

11. Order of Works   Employer's requirements affecting the order of the
Main Contract Works (if any).

12. Location and
type of access

ARCHITECT - STAGE 2

*If extensive this information
could be given by a copy of
the Preliminary Bill of the
Main Contract Bills

13.* Obligations or
restrictions
imposed by
Employer not
covered by
Main Contract
Conditions
(e.g. in the
Preliminaries in
the Contract Bills)

14.* Other relevant
information
(if any)

We have noted the foregoing information

NOMINATED SUB-CONTRACTOR

—STAGE 3

_____Sub-Contractor.

Page 5

205

Fluctuations (see Page 3 item 2)

N.B. – This Appendix should be completed by the Sub-Contractor where
Sub-Contract NSC/4 Clauses 35 or 36 apply, attaching further sheets if necessary.

Where Clause 35 applies only column 1 need be completed.
Where Clause 36 applies columns 1 to 4 must be completed.

The 'Date of Tender' for the purpose of Clauses 35 and 36 is ~~ARCHITECT~~ _STAGE 2_ ___19___

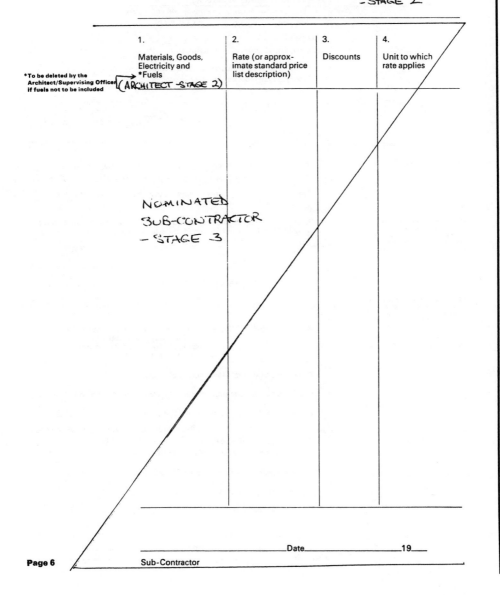

1. Materials, Goods, Electricity and *Fuels	2. Rate (or approximate standard price list description)	3. Discounts	4. Unit to which rate applies

*To be deleted by the Architect/Supervising Officer if fuels not to be included → (ARCHITECT -STAGE 2)

NOMINATED SUB-CONTRACTOR - STAGE 3

_____ Date _____ 19___

Page 6   Sub-Contractor

Fluctuations (see Page 3 item 2)

N.B.–This Appendix should be completed by the Sub-Contractor when Sub-Contract NSC/4 Clause 37 applies.

ARCHITECT – STAGE 2

*Delete as applicable

(i)   37.1 – Nominated Sub-Contract Formula Rules are those dated _____ 19__
Part I*/ Part III of these Rules applies.

*Not to exceed 10%

(ii)   37.3.3 and 37.3.4 *Non-Adjustable Element _____%
(Local Authority with quantities only).

(iii)   37.4 – List of Market Prices *(use separate sheet if required).*

NOMINATED SUB-CONTRACTOR – STAGE 3

(iv)   Nominated Sub-Contract Formula Rules.

Rule 3. (Definition of Balance of Adjustable Work).
Any measured work not allocated to a Work Category.

Base Month _____ 19___   ARCHITECT – STAGE 2

Date of Tender _____ 19___

Rule 8.

*Delete as applicable

Method of dealing with 'Fix only' work.   Rule 8(i)*/8(ii)*/8(iii) applies

Rule 11(a).   NOMINATED SUB-CONTRACTOR – STAGE 3

Part I only: the Work Categories applicable to the Sub-Contract Works.

Rule 43.

If both specialist engineering formulae apply to the Sub-Contract the percentages for use with each formula should be inserted and clearly identified

Part III only: Weightings of labour and materials – Electrical Installations or Heating, Ventilating and Air Conditioning Installations.

ARCHITECT – STAGE 2

	Labour	Materials
Electrical	_____%	_____%
*Heating, Ventilating and Air Conditioning	_____%	_____%
Sprinklers	_____%	_____%

*The weightings for sprinkler installations may be inserted where different weightings are required

Rule 61a.

*Delete as applicable

Adjustment shall be effected *upon completion of manufacture of all fabricated components.
*upon delivery to site of all fabricated components.

Rule 64.

Part III only Structural Steelwork Installations:

(i)   Average price per tonne of steel delivered to fabricator's works £_____

(ii)   Average price per tonne for erection of steelwork £_____

Rule 70a.   NOMINATED SUB-CONTRACTOR – STAGE 3

Catering Equipment Installations:
Apportionment of the values of each item between

(i)   Materials and shop fabrication £_____

(ii)   Supply of factor items   £_____

(iii)   Site installations   £_____

_____ Date _____ 19___
Sub-Contractor.

　　　Schedule 2

Schedule 2: Particular Conditions

Note: When the Contractor receives Tender NSC/1/Scot together with the Architect/Supervising Officer's preliminary notice of nomination under Clause 35.7.1 of the Main Contract Conditions then the Contractor has to agree this Schedule with the proposed Sub-Contractor.

To be completed by the Architect/Supervising Officer before issue

1.A. Any stipulations as to the period/periods when Sub-Contract Works to be

carried out on site to be between

and　　ARCHITECT – STAGE 2

Period required by Architect to approve drawings after submission _____ weeks

The Sub-Contractor will set out here (or on an attached sheet if necessary) details of the carrying out of the Sub-Contract Works as a preliminary indication to the Architect and Contractor. Adaptation will be needed where a phased completion is required

*Not including any period for approval

1.B. Preliminary programme
(having regard to the information provided in the invitation to tender).
Period required:

*for preparation of further design work and/or drawings _____ weeks

*for preparation of working drawings or installation drawings _____ weeks

*for preparation of shop drawings (if required) _____ weeks

for execution of the Sub-Contract Works off-site _____ weeks

NOMINATED SUB-CONTRACTOR
STAGE 3　　　　　　　　　　on-site _____ weeks

Notice required to commence work on site _____ weeks

Agreed details on the programme for carrying out the Sub-Contract Works (which must include the subjects set out in 1.A and 1.B) must be inserted here or on an attached sheet initialled by the Contractor and Sub-Contractor. The details of 1.A and 1.B (and any sheet attached thereto) must then be deleted and the deletion initialled by the Contractor and the Sub-Contractor

1.C. Agreed programme details (including sub-contract completion date: see Sub-Contract NSC/4 Clause 11.1).

BY AGREEMENT
CONTRACTOR AND NOMINATED
SUB-CONTRACTOR – STAGE 7

2. Order of Works – to follow the requirements, if any, stated in Schedule 1, item 11.

208

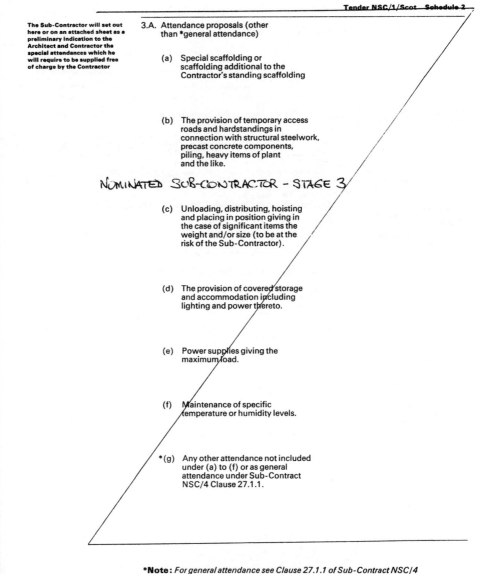

The Sub-Contractor will set out here or on an attached sheet as a preliminary indication to the Architect and Contractor the special attendances which he will require to be supplied free of charge by the Contractor

3.A. Attendance proposals (other than *general attendance)

(a) Special scaffolding or scaffolding additional to the Contractor's standing scaffolding

(b) The provision of temporary access roads and hardstandings in connection with structural steelwork, precast concrete components, piling, heavy items of plant and the like.

NOMINATED SUB-CONTRACTOR - STAGE 3

(c) Unloading, distributing, hoisting and placing in position giving in the case of significant items the weight and/or size (to be at the risk of the Sub-Contractor).

(d) The provision of covered storage and accommodation including lighting and power thereto.

(e) Power supplies giving the maximum load.

(f) Maintenance of specific temperature or humidity levels.

*(g) Any other attendance not included under (a) to (f) or as general attendance under Sub-Contract NSC/4 Clause 27.1.1.

*Note: *For general attendance see Clause 27.1.1 of Sub-Contract NSC/4 which states:*

'General attendance shall be provided by the Contractor free of charge to the Sub-Contractor and shall be deemed to include only use of the Contractor's temporary roads, pavings and paths standing scaffolding, standing power operated hoisting plant, the provision of temporary lighting and water supplies, clearing away rubbish, provision of space for the Sub-Contractor's own offices and for the storage of his plant and materials and the use of mess rooms; sanitary accommodation and welfare facilities: see S.M.M. 6th Edition B.9.2.

Page 9

Tender NSC/1/Scot   Schedule 2

The Contractor will set out in
agreement with the Sub-
Contractor any alterations to any
of the details of the attendance
set out in 3A above

3.B.   Agreed special attendances.

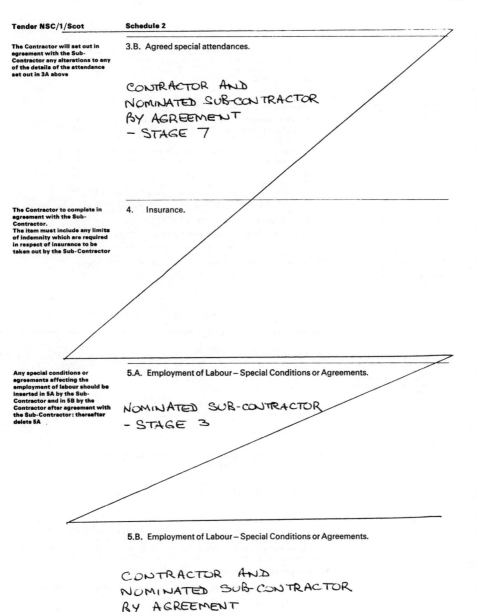

CONTRACTOR AND
NOMINATED SUB-CONTRACTOR
BY AGREEMENT
– STAGE 7

The Contractor to complete in
agreement with the Sub-
Contractor.
The item must include any limits
of indemnity which are required
in respect of insurance to be
taken out by the Sub-Contractor

4.   Insurance.

Any special conditions or
agreements affecting the
employment of labour should be
inserted in 5A by the Sub-
Contractor and in 5B by the
Contractor after agreement with
the Sub-Contractor: thereafter
delete 5A .

5.A.   Employment of Labour – Special Conditions or Agreements.

NOMINATED SUB-CONTRACTOR
– STAGE 3

5.B.   Employment of Labour – Special Conditions or Agreements.

CONTRACTOR AND
NOMINATED SUB-CONTRACTOR
BY AGREEMENT
– STAGE 7

210

6. *This item which deals with the appointment of an Adjudicator and Trustee Stakeholder is not applicable to Scotland but it is retained to maintain the same numerical sequence as in NSC/1/Schedule 2 published by the J.C.T. for use in England.*

To be completed by the Contractor after agreement with the Sub-Contractor

7. Finance (No. 2) Act 1975 – Statutory Tax Deduction Scheme.

    1. The Contractor is/is not entitled to be paid by the Employer without the statutory deduction referred to in the above Act or such other deduction as may be in force;

    2. The Sub-Contractor is/is not entitled to be paid by the Contractor without the above-mentioned statutory deduction or such other deduction;

    3. The evidence to be produced to the Contractor for the verification of the Sub-Contractor's tax certificate (expiry date   19   ) will be: *CONTRACTOR AND NOMINATED SUB-CONTRACTOR BY AGREEMENT – STAGE 7*

*Delete as applicable

8. Value Added Tax – Sub-Contract NSC/4.

    Clause 19A*/19B (alternative provisions) will apply.

The Sub-Contractor will set out here or on an attached sheet any matter he wishes to agree with the Contractor

9.A. Any other matters (including any limitation on working hours).

NOMINATED SUB-CONTRACTOR – STAGE 3

To be completed by the Contractor after agreement with the Sub-Contractor. Thereafter 9A should be deleted and initialled by the Contractor and Sub-Contractor

9.B.

CONTRACTOR AND NOMINATED SUB-CONTRACTOR BY AGREEMENT – STAGE 7

10. Any other matters agreed by the Architect/~~Supervising Officer~~ and Sub-Contractor before preliminary notice of nomination.

ARCHITECT – STAGE 6

11. *This item which indicates whether the Sub-Contract is or is not executed under Seal is not applicable to Scotland but it is retained to maintain the same numerical sequence as in NSC/1 Schedule 2 published by the J.C.T. in England.*

The above Particular Conditions are agreed

NOMINATED SUB-CONTRACTOR – STAGE 7

_____Date_____19____

Signed by or on behalf of the Sub-Contractor.

CONTRACTOR – STAGE 7

_____Date_____19____

Signed by or on behalf of the Main Contractor.

Page 11

Explanatory notes

Although Tender NSC/1/Scot is a white document, it is virtually the same as its English counterpart NSC/1, having the same clauses and numbering throughout. The format differs and care must be taken with the entries.

The following clauses are deleted in order to bring the signing of contracts into line with Scottish law and practice (contracts are not executed under seal in Scotland):

Schedule 1   Item 7
Schedule 2   Item 11

Again, Item 6 of Schedule 2 is omitted since it deals with the appointment of an adjudicator and trustee stakeholder, and this is not applicable in Scotland.

With the above exceptions, the explanatory notes for Tender NSC/1 in England and Wales hold good for the Scottish version, NSC/1/Scot.

N.B.

April 1981 Revision

As compared with the original the following alterations have been made.

1. Clause 2 – re-written to comply with identical clause in English NSC/2.

2. Clause 9 – Addition to end of clause excepting 35.24.6.

Agreement NSC/2/Scot
(Revised April 1981)

AGREEMENT

between

(Employer)

and

(Sub-Contractor)

This form is issued by the
Scottish Building Contract Committee of which the
constituent bodies are:

Royal Incorporation of Architects in Scotland
Scottish Building Employers Federation
Scottish Branch of the Royal Institution of Chartered Surveyors
Convention of Scottish Local Authorities
Federation of Specialists and Sub-Contractors
    (Scottish Board)
Committee of Associations of Specialist Engineering
    Contractors (Scottish Branch)
Association of Consulting Engineers (Scottish Group)
Confederation of British Industry
Association of Scottish Chambers of Commerce

———————

Copyright of the S.B.C.C., 39 Castle Street, Edinburgh

April 1981

Agreement NSC/2/Scot
(Revised April 1981)

Agreement between an Employer and a Sub-Contractor nominated under Clauses 35.6 - 35.10 of the Standard Form of Building Contract

Main Contract Works:

Location:

Sub-Contract Works:

ARCHITECT - STAGE 2   **AGREEMENT**

between

(hereinafter referred to as 'the Employer')

and

(hereinafter referred to as 'the Sub-Contractor')

*Insert here the date of
Sub-Contractor's offer on
Page 1 of NSC/1/Scot

CONSIDERING THAT the Sub-Contractor has submitted a tender dated*_____
on Tender NSC/1/Scot (hereinafter referred to as 'the Tender') to carry out certain Works
(hereinafter referred to as 'the Sub-Contract Works') forming part of the Main Contract
Works being or to be carried out on the terms and conditions set out in Schedule 1
to the Tender:
FURTHER CONSIDERING that the Employer has appointed

to be the Architect/Supervising Officer for the purposes of the Main Contract and this
Agreement (who, and his validly appointed successors, is hereinafter referred to as
'the Architect/Supervising Officer'):
FURTHER CONSIDERING that the Architect/Supervising Officer has approved the
Tender on behalf of the Employer and proposes to nominate the Sub-Contractor on
Nomination NSC/3/Scot to carry out and complete the Sub-Contract Works on the
terms and conditions referred to in the Tender after the Sub-Contractor and the
Main Contractor referred to in the Tender have agreed on the Particular Conditions
set out in Schedule 2 to the Tender

THEREFORE the Employer and the Sub-Contractor HAVE AGREED and DO HEREBY AGREE as follows:

1 The Sub-Contractor shall, after the Architect/Supervising Officer has issued his preliminary notice of nomination, forthwith agree with the Main Contractor the said Particular Conditions and shall immediately inform the Architect/Supervising Officer when agreement has been reached and Schedule 2 has been signed by him and the Contractor.

2 The Sub-Contractor warrants that he has exercised and will exercise all reasonable skill and care in

2.1 the design of the Sub-Contract Works in so far as the Sub-Contract Works have been or will be designed by him,

2.2 the selection of materials and goods for the Sub-Contract Works in so far as such materials and goods have been or will be selected by him, and

2.3 the satisfaction of any performance specification or requirement in so far as such performance specification or requirement is included or referred to in the description of the Sub-Contract Works included in or annexed to the Tender.

3 If after execution of this Agreement and before the issue by the Architect/Supervising Officer of the nomination instruction Nomination NSC/3/Scot the Architect/Supervising Officer instructs the Sub-Contractor in writing to proceed with the designing of or the proper ordering or fabrication of any materials or goods for the Sub-Contract Works the Sub-Contractor shall forthwith comply with the instruction and

3.1 the Employer shall pay the Sub-Contractor the amount of any expense reasonably and properly incurred by him in carrying out work in the designing of the Sub-Contract Works and upon such payment the Employer may use that design work for the purposes of the Sub-Contract Works but not otherwise;

3.2 the Employer shall pay the Sub-Contractor for any materials or goods properly ordered or fabricated by the Sub-Contractor for the Sub-Contract Works and upon such payment any materials and goods shall become the property of the Employer;

3.3 the Employer shall not be required to make any payment under Clauses 3.1 and 3.2 hereof after the issue of Nomination NSC/3/Scot by the Architect/Supervising Officer except in respect of any design work properly carried out and/or materials or goods properly ordered or fabricated in compliance with a prior instruction under Clause 3 hereof but which are not used for the Sub-Contract Works by reason of some written decision of the Architect/Supervising Officer before the issue of Nomination NSC/3/Scot;

3.4 if the Employer has made any payment under Clauses 3.1 or 3.2 hereof and the Sub-Contractor is subsequently nominated to execute the Sub-Contract Works in accordance with Nomination NSC/3/Scot the Sub-Contractor shall allow the Employer and the Main Contractor full credit for such payment in the discharge of the amounts due in respect of the Sub-Contract Works.

4 The Sub-Contractor shall supply the Architect/Supervising Officer with such information (including drawings) in accordance with the agreed programme details or at such time as the Architect/Supervising Officer may reasonably require so that the Architect/Supervising Officer shall not be delayed in issuing necessary instructions or drawings under the Main Contract for which delay the Contractor might have a valid claim for an extension of time for completion of the Main Contract Works under Clause 25.4.6 or a valid claim for direct loss and/or expense under Clause 26.2.1 of the Main Contract Conditions.

5 The Sub-Contractor shall perform the Sub-Contract so that the Architect/Supervising Officer will not by reason of any default of the Sub-Contractor be required to issue an instruction to determine the employment of the Sub-Contractor under the Main Contract: Provided that any suspension by the Sub-Contractor of further execution of the Sub-Contract Works under Clause 21.8 of Sub-Contract NSC/4 shall not be regarded as coming within this clause.

6 The Sub-Contractor shall perform the Sub-Contract so that the Contractor will not become entitled to an extension of time for completion of the Main Contract Works.

7 The Employer agrees that the Sub-Contractor will not be liable under Clauses 4, 5 and 6 hereof until the Architect/Supervising Officer has issued Nomination NSC/3/Scot.

8 The Employer shall ensure that the Architect/Supervising Officer will operate the provisions of Clauses 35.13.1 * and 35.17† to 35.19 of the Main Contract Conditions.

* Clause 35.13.1 requires the Architect/Supervising Officer when directing the Contractor as to the amount included in an Interim Certificate in respect of the value of the Nominated Sub-Contract Works issued under Clause 30 of the Main Contract Conditions forthwith to inform the Sub-Contractor in writing of that amount

† Clause 35.17 deals with final payment by the Employer for the Sub-Contract Works prior to the issue of the Final Certificate under the Main Contract Conditions

8.1 After final payment to the Sub-Contractor under Clause 35.17 of the Main Contract Conditions the Sub-Contractor shall rectify at his own cost (or if he fails to do so shall be liable to the Employer for the costs referred to in Clause 35.18 of the Main Contract Conditions) any omission, fault or defect in the Sub-Contract Works which the Sub-Contractor is bound in accordance with the Sub-Contract to rectify after written notification of the same by the Architect/Supervising Officer at any time prior to the issue of the Final Certificate under the Main Contract.

8.2 After the issue of the said Final Certificate the Sub-Contractor shall in addition to such other responsibilities, if any, as he has under this Agreement, have the same responsibility to the Contractor and to the Employer for the Sub-Contract Works as the Contractor has to the Employer under the Main Contract after the issue of the Final Certificate.

9 Where the Architect/Supervising Officer has been required under Clause 35.24 of the Main Contract Conditions to make a further nomination in respect of the Sub-Contract Works, the Sub-Contractor shall indemnify the Employer against any direct loss and/or expense resulting therefrom, except where Clause 35.24.6 applies.

10 The Employer shall operate, and shall ensure that the Architect/Supervising Officer shall operate, the provisions in regard to the payment of the Sub-Contractor contained in Clause 35.13 of the Main Contract Conditions.

10.1 If, after paying any amount to the Sub-Contractor under Clause 35.13.5.3 of the Main Contract Conditions, the Employer produces reasonable proof that there was in existence at the time of such payment a petition or resolution to which Clause 35.13.5.4.4 of the Main Contract Conditions refers, the Sub-Contractor shall repay such amount on demand.

*Clause 2.3 deals with specified supplies and restrictions, etc. in the contracts of sale for such supplies

11 Where Clause 2.3* of Sub-Contract NSC/4 applies the Sub-Contractor shall forthwith give the Contractor details of any restriction, limitation or exclusion to which that clause refers as soon as such details are known to the Sub-Contractor.

12 If any conflict appears between the terms of the Tender and this Agreement, the terms of this Agreement shall prevail.

13 In the event of any dispute or difference of opinion arising regarding any term of this Agreement or as to any matter or thing arising hereunder or in connection herewith the same shall be and is hereby referred to the decision of an Arbiter mutually agreed on, or failing such agreement within 14 days after either party has given written notice to the other to concur in the appointment of an Arbiter, by an Arbiter appointed by the Sheriff of any Sheriffdom in which the Sub-Contract Works or any part thereof are situated.

13.1 The Arbiter shall without prejudice to the generality of his powers have power to make such orders and directions as may in his opinion be necessary to determine the rights of the parties and to open up review and revise any directions, requirements or notices given by either party to the other, and to determine all matters in dispute as if no such direction, requirement or notice had been given.

13.2 The Law of Scotland shall apply to all arbitrations under these presents and the award of the Arbiter shall be final and binding on the parties subject always to the provisions of Section 3 of the Administration of Justice (Scotland) Act 1972.

13.3 The Arbiter shall be entitled to remuneration and reimbursement of his outlays.

14 This Agreement shall be construed and the rights of the Employer and the Sub-Contractor and all matters arising hereunder shall be determined according to the Law of Scotland: IN WITNESS WHEREOF these presents are executed at   on the   day of   19   before these witnesses subscribing.

_____witness

_____address

_____occupation

EMPLOYER - STAGE 5

_____Employer.

_____witness

_____address

_____occupation

_____witness

_____address

_____occupation

NOMINATED SUB-CONTRACTOR - STAGE 3

_____Sub-Contractor.

_____witness

_____address

_____occupation

N.B.–This document is set out for execution by individuals or firms; when limited companies or local authorities are involved amendment will be necessary and the appropriate officials should be consulted.   **Page 4**

Explanatory notes

Again, Agreement NSC/2/Scot is a white document, it does not have recitals as such, but the same information is given in a slightly different format. The clause numbering is different and therefore they are in a different order.

The following clauses are changed for Scotland:

Clause 13   For arbitration there is appointed an 'arbiter' (rather than the English 'arbitrator'), and if mutual agreement to appoint an arbiter within 14 days is not forthcoming, in Scotland the Sheriff of the Sheriffdom in which the Sub-contract Works are situated appoints an arbiter.

Clause 13.2   Here there is no choice given for amendment – the law of Scotland will apply.

Clause 13.3   This is additional, setting out an arbiter's right to recover his fees and expenses.

Although the clauses are in a different order (sometimes two sub-clauses becoming one, the numbering not the same and changes as noted above), the explanatory notes for Agreement NSC/2 in England and Wales can be read in conjunction with this Scottish version since the meanings and intentions of the clauses remain the same.

218

Standard Form for Nomination of Sub-Contractors for use in Scotland
(where Tender NSC/1/Scot has been used)

To (Main Contractor)

Main Contract Works

ARCHITECT

Sub-Contract Works *STAGE 8*

Page(s) number of Bills
or Specification

Name and address of
Nominated Sub-Contractor

Further to   (1) My/Our preliminary notice of nomination dated_____

and (2) The Tender NSC/1/Scot and Schedules thereto completed by the
Sub-Contractor referred to above and approved by me/us on
behalf of the Employer

I/we hereby nominate the Sub-Contractor referred to above under Clause 35.10.2
of the Main Contract.

_____
Architect/Supervising Officer

Date_____

Issued to	Main Contractor	Q.S.	Nominated Sub-Contractor	Clerk of Works	Consulting Engineer	A/S.O. file
	☐	☐	☐	☐	☐	☐

Copyright SBCC, 39 Castle Street, Edinburgh   February 1980

Explanatory notes

Again the form is white-coloured and although the layout is not exactly the same, there is no real difference from the English/Welsh version. Accordingly, the explanatory notes for Nomination NSC/3 in England and Wales can be read in conjunction with this Scottish version, NSC/3/Scot.

N.B.

February 1983 Revision.

As compared with the original, the following alterations have been made:

1. Sub-Contract

    P.1 Second Paragraph
        word in brackets altered to comply with Instructions for Use in NSC/1/Scot

    P.2 Clause 3.1.1
        'Payment' substituted for 'certificate'

        Clause 3.1.4
        '4.3' substituted for '4.6'

    P.4 21
        Alteration to clause 21.3.2.1 added

        21.4.4 'or Tender Sum' added after 'Sub-Contract Sum'

Scottish Building Sub-Contract NSC/4/Scot
(Revised February 1983)

BUILDING SUB-CONTRACT

between

and

with

SCOTTISH SUPPLEMENT 1980

to

The Conditions of the Standard Form of Sub-Contract
for Sub-Contractors nominated under Clause 35.10.2 of the
Standard Form of Building Contract
1980 Edition

The constituent bodies of the
Scottish Building Contract Committee are:

Royal Incorporation of Architects in Scotland
Scottish Building Employers Federation
Scottish Branch of the Royal Institution of Chartered Surveyors
Convention of Scottish Local Authorities
Federation of Specialists and Sub-Contractors
    (Scottish Board)
Committee of Associations of Specialist Engineering
    Contractors (Scottish Branch)
Association of Consulting Engineers (Scottish Group)
Confederation of British Industry
Association of Scottish Chambers of Commerce

Copyright of the S.B.C.C., 39 Castle Street, Edinburgh

February 1983

Scottish Building Sub-Contract NSC/4/Scot
(Revised February 1983)

For use in Scotland for Sub-Contractors who have tendered
on Tender NSC/1/Scot, executed Agreement NSC/2/Scot
and have been nominated by Nomination NSC/3/Scot under
the Standard Form of Building Contract Clause 35.10.2.

BUILDING SUB-CONTRACT

between

(hereinafter referred to as 'the Contractor')

CONTRACTOR
OR ARCHITECT
STAGE 9

and

(hereinafter referred to as 'the Sub-Contractor')

———————

CONSIDERING THAT the Sub-Contractor has submitted a tender on Tender NSC/1/Scot offering
to carry out and complete certain Sub-Contract Works forming part of a Main Contract being
carried out by the Contractor, all as the said Sub-Contract Works and Main Contract are described
in the said Tender:

FURTHER CONSIDERING that the said Tender has been approved by the Architect therein
referred to and accepted by the Contractor (and the original has been retained by the Contractor
and copies by the Employer in the Main Contract and the Sub-Contractor):

FURTHER CONSIDERING that the Sub-Contractor has been given notice of the provisions of the
Main Contract except the detailed prices of the Contractor included in schedules and bills of
quantities:

FURTHER CONSIDERING that the Sub-Contractor has been nominated by the Architect on
Nomination NSC/3/Scot to carry out the Sub-Contract Works:

FURTHER CONSIDERING that at the date of this Building Sub-Contract

*Delete as applicable

(a) the Sub-Contractor is*/is not the user of a current Sub-Contractor's Tax Certificate under the
provisions of the Finance (No. 2) Act 1975 in one of the forms specified in Regulation 15 of the
Income Tax (Sub-Contractors in the Construction Industry) Regulations 1975 and the Schedule
thereto:

(b) the Contractor is*/is not the user of a current Sub-Contractors Tax Certificate under the said Act
and Regulations; and

(c) the Employer under the Main Contract is*/is not 'a Contractor' within the meaning of the said Act
and Regulations:

THEREFORE the Contractor and the Sub-Contractor HAVE AGREED and DO HEREBY AGREE

*Complete in accordance
with Tender NSC/1/Scot

1 The Sub-Contractor will carry out and complete the Sub-Contract Works for
*the sum stated in the said Tender or such other sum as shall become payable in accordance with
this Building Sub-Contract.

OR

*the Tender Sum stated in the said Tender as finally ascertained in accordance with this
Building Sub-Contract (hereinafter referred to as 'the Ascertained Final Sub-Contract Sum').

2

2 The Sub-Contract Works shall be carried out and completed in accordance with, and the rights and duties of the Contractor and Sub-Contractor shall be regulated by Conditions of the Standard Form of Sub-Contract for Sub-Contractors nominated under the Standard Form of Building Contract (known as Sub-Contract NSC/4) issued by the Joint Contracts Tribunal which are held to be incorporated in and form part of this Sub-Contract subject only to the amendments and modifications contained in the Scottish Supplement forming the Appendix hereto.

3 In the event of any dispute or difference between the Contractor and the Sub-Contractor arising during the progress of the Sub-Contract Works or after completion or abandonment thereof in regard to any matter or thing whatsoever arising out of this Sub-Contract or in connection herewith, then such dispute or difference shall be and is hereby referred to the arbitration of such person as the parties may agree to appoint as Arbiter or failing agreement within 14 days after either party has given to the other written notice to concur in the appointment of an Arbiter as may be appointed by the Sheriff of any Sheriffdom in which the Sub-Contract Works or any part thereof are situated: Arbitration proceedings shall be deemed to have been instituted on the date on which the said written notice has been given.

3.1 No arbitration shall commence without the written consent of the parties until after determination or alleged determination of the Contractor's employment under the Main Contract or until after Practical Completion or alleged Practical Completion of the Main Contract Works or abandonment of the Main Contract Works unless it relates to

   3.1.1 whether or not a payment has been improperly withheld or is not in accordance with the said Conditions.

   3.1.2 whether Practical Completion of the Sub-Contract Works has taken place in accordance with Clause 14.2.

   3.1.3 a claim by the Contractor or counter-claim by the Sub-Contractor under Clause 24.

   3.1.4 a dispute under Clauses 4.3, 11.2 or 11.3.

3.2 If the dispute or difference is substantially the same as or is connected with a dispute or difference between the Contractor and Employer under the Main Contract the Contractor and Sub-Contractor hereby agree that such dispute or difference shall be referred to an Arbiter appointed or to be appointed to determine the related dispute or difference provided that either party may require the appointment of a different Arbiter if he reasonably considers the Arbiter appointed in the related dispute is not suitably qualified to determine the dispute or difference under this Sub-Contract.

3.3 The Arbiter shall have power to

   3.3.1 direct such measurements and/or valuations as may in his opinion be desirable in order to determine the rights of the parties

   3.3.2 ascertain and award any sum which ought to have been referred to or included in any certificate

   3.3.3 open up review and revise any certificate, opinion, decision, requirement or notice

   3.3.4 determine all matters in dispute which shall be submitted to him in the same manner as if no such certificate, opinion, decision, requirement or notice had been given

   3.3.5 award compensation or damages and expenses to or against any of the parties to the arbitration.

3.4 In any arbitration under this clause a decision of the Architect which is final and binding on the Contractor shall also be final and binding between and upon the Contractor and the Sub-Contractor.

3.5 The Law of Scotland shall apply to all arbitrations in terms of this clause and the award of the Arbiter shall be final and binding on the parties subject to the provisions of ~~Clause~~ *Section* 3 of the Administration of Justice (Scotland) Act 1972.

3.6 The Arbiter shall be entitled to remuneration and reimbursement of his outlays.

3

4   Both parties agree that this Sub-Contract shall be regarded as a Scottish Contract and shall be construed and the rights of parties and all matters arising hereunder determined in all respects according to the Law of Scotland.

5   Both parties consent to registration hereof for preservation and execution:

IN WITNESS WHEREOF these presents are executed as follows:

Signed by the above named Contractor

on the   day of   19   before these

witnesses

_____witness

_____address

_____occupation   *CONTRACTOR*
                                             - STAGE 9
                                             _____ Contractor
                                             (Attention is drawn to the note
                                             at the foot of this page)

_____witness

_____address

_____occupation

Signed by the above named Sub-Contractor

on the   day of   19   before these

witnesses

_____witness

_____address

_____occupation   *SUB-CONTRACTOR*
                                             - STAGE 9
                                             _____ Sub-Contractor
                                             (Attention is drawn to the note
                                             at the foot of this page)

_____witness

_____address

_____occupation

N.B. – This document is set out as for execution by individuals or firms: Where Limited Companies are involved amendment will be necessary and the appropriate officials should be consulted.

Both parties sign here and on page 5.

4

SCOTTISH SUPPLEMENT

(The following are the amendments and modifications to the
Conditions of Standard Form of Sub-Contract NSC/4.
The numbers refer to clauses in NSC/4).

1   Interpretation, Definitions, etc.

1.1 and 1.2 shall be deleted.

1.3   The meanings given to the undernoted words and phrases shall be deleted and the
following substituted:

Agreement NSC/2	Agreement NSC/2/Scot
Arbitrator	Arbiter
Article or Articles of Sub-Contract Agreement	The foregoing Building Sub-Contract
Sub-Contract NSC/4	Building Sub-Contract NSC/4/Scot
Tender	A duly completed and signed Tender NSC/1/Scot

The following clause shall be added:

1.4   Additional definition
Real or personal   Heritable or moveable

19A and 19B – Value Added Tax

Delete as applicable   *19A shall apply

*19B shall apply.

N.B. (1) Clause 19B applies only when the Main Contractor under the VAT (General) Regulations
1972 (Regulation 8(3)) has with the approval of the Sub-Contractor been allowed to prepare
tax documents in substitution for an authenticated receipt issued by the Sub-Contractor under
Regulation 21(2).

(2) Where Clause 19B applies Clause 19A must be deleted.

20A and 20B – Finance (No. 2) Act 1977 – Tax Deduction Scheme

Delete as applicable   *20A shall apply.

*20B shall apply.

N.B. Clause 20A applies when the Sub-Contractor has a current Sub-Contractor's Tax
Certificate.

Clause 20B applies when the Sub-Contractor has no such Certificate. See narrative of the
foregoing Building Sub-Contract.

20A.3.2/20B.3 shall be deleted and the following substituted:

The Contractor shall immediately inform the Sub-Contractor of any change in his own
status or that of the Employer as stated in the foregoing Building Sub-Contract or
Tender NSC/1/Scot (Schedule 2).

20A.8/20B.6 There shall be deleted 'the provisions of Article 3 shall apply to' and added at end
'shall be referred to an Arbiter appointed in accordance with the foregoing Building
Sub-Contract.'

21   Payment of Sub-Contractor

21.2.3 shall be deleted.

21.3.2.1 line 4 Delete the words 'Clause 5' and substitute the words 'Clause 8.'

21.4.1.3 shall be deleted.

The following clause shall be added:

21.4.4 If the Architect/Supervising Officer is of the opinion that it is expedient to do so, the
Sub-Contractor may with the consent of the Contractor, which consent shall not be
unreasonably withheld, enter into a Contract for the purchase of materials and/or goods
by the Employer in the Main Contract prior to their delivery to the site, and upon such
Contract being entered into the purchase of the said off-site materials and/or goods shall
be excluded altogether from this Sub-Contract and the Sub-Contract sum or Tender sum
shall be adjusted accordingly: And it is specifically declared and provided that payment
by the Employer to the Sub-Contractor for any of the said materials and/or goods shall in
no way affect any cash discounts or other emoluments to which the Contractor may be
entitled.

5

24   Contractors claims not agreed by the Sub-Contractor – appointment of Adjudicator

24.1.2 shall be deleted and the following substituted:

Subject to the provisions of Clauses 21.3, 23 and 24 hereof the Sub-Contractor shall be entitled to appoint an Adjudicator to be selected from a list maintained by the Scottish Building Contract Committee to decide those matters referable to the Adjudicator under the provisions of Clause 24. In the event of the first person approached declining to act the Sub-Contractor shall approach another person on the said list and so on until the appointment is made: Provided that no Adjudicator shall be appointed who has any interest in this Sub-Contract or the Main Contract of which this Sub-Contract is a part or in other Contracts or Sub-Contracts in which the Contractor or the Sub-Contractor is engaged, unless the Contractor, Sub-Contractor and Adjudicator so mutually agree in writing.

24.3.1.2 shall be deleted and the following substituted:

shall be placed on Deposit Receipt in joint names of the Contractor and Sub-Contractor with such Bank as the Adjudicator shall direct pending the result of the arbitration.

24.4.1 shall be deleted and the following substituted:

Where any decision of the Adjudicator notified under Clause 24.3.3 requires the Contractor to place an amount on joint Deposit Receipt as aforesaid the Contractor shall forthwith do so and shall send a duplicate of the said Deposit Receipt to the Sub-Contractor immediately on issue: Provided that the Contractor shall not be obliged to deposit a sum greater than the amount due from the Contractor under Clause 21.3 hereof in respect of which the Contractor has exercised the right of set-off referred to in Clause 23.2 hereof.

24.5 shall be deleted and the following substituted:

Any amount placed on joint Deposit Receipt as aforesaid shall remain on deposit until the Arbiter appointed pursuant to the aforesaid notice of arbitration directs that it should be uplifted or the Contractor and Sub-Contractor mutually so agree. In either case the said Deposit Receipt shall be endorsed by the Contractor and the Sub-Contractor and the sum on Deposit Receipt (with interest if appropriate) uplifted and paid to the Contractor or the Sub-Contractor in accordance with the Arbiter's award or as the Contractor and Sub-Contractor may have mutually agreed as the case may be.

24.8 There shall be deleted 'and where relevant, for the charges of the Trustee-Stakeholder or any part thereof.'

29   Determination of the employment of the Sub-Contractor by the Contractor

29.2 shall be deleted and the following substituted:

In the event of a Provisional Liquidator being appointed to control the affairs of the Sub-Contractor the Contractor may determine the employment of the Sub-Contractor under this Contract by giving him 7 days written notice sent by Registered Post or Recorded Delivery of such determination. In the event of the Sub-Contractor becoming bankrupt or making a composition or arrangement with his creditors or having his estate sequestrated or being rendered notour bankrupt or entering into a Trust Deed with his creditors or having a winding up order made or (except for the purpose of reconstruction) a resolution for voluntary winding up passed or a Receiver or Manager of his business or undertaking duly appointed or possession taken by or on behalf of the holders of any Debenture secured by a Floating Charge then without prejudice to any other rights or remedies which the Contractor may have the employment of the Sub-Contractor under this Sub-Contract shall be automatically determined.

29.3.3 The words 'any such property of the Sub-Contractor' shall be deleted and the words 'any such property so far as belonging to the Sub-Contractor' substituted.

31   Determination of the Main Contractor's employment under the Main Contract

The following clause shall be added:

31.3 The Sub-Contractor shall recognise an assignation by the Contractor in favour of the Employer in terms of Clause 27.4.2.1 of the Main Contract Conditions.

STAGE 9
_____Contractor.

STAGE 9
_____Sub-Contractor.

Explanatory notes

Both England-and-Wales and Scotland use the same set of Sub-contract Conditions, namely NSC/4 for the Basic Method.

In Scotland it is a Building Sub-contract that is completed by the Contractor and the Nominated Sub-contractor, not Articles of Agreement as in England and Wales. Again, differing from South of the border, this white form is detached from the Sub-contract Conditions, which themselves are incorporated only by reference. Also in the form is the Scottish Supplement to allow the necessary changes to be incorporated in the Sub-contract Conditions so as, to enable Scottish Law and practice to operate.

Although both Contractor and Nominated Sub-contractor have copies of the completed Scottish Building Sub-contract Form NSC/4/Scot, it is still necessary to have a separate copy of Sub-contract Conditions NSC/4 for reference.

Scottish Building Sub-contract

On the first page in paragraph 2, in brackets at the end it is stated that the original or principal copy of Tender NSC/1/Scot be retained by the Employer. *Diagram 1(S)*, (page 95), suggests that it be passed to the Contractor for safe keeping and presumeably attached to the principal NSC/4/Scot. This is in accordance with the instructions on the back page of Tender NSC/1/Scot. Accordingly, if this is done then the words in brackets would need to be deleted and initialled by both parties.

In the fifth paragraph it is necessary to state, by deletion, whether or not each party is a contractor in the terms of the Finance (No 2) Act 1975. These entries have already been made in Tender NSC/1/Scot and, unless any change in circumstances has taken place, they will be copied here.

Clause 1   The alternative here will be copied from Tender NSC/1/Scot and care must be taken to ensure that the correct amount is placed in the appropriate alternative.

Clause 2   This is where the reference to Sub-contract Conditions NSC/4 is made, and therefore incorporates it into Sub-contract NSC/4/Scot. Note that *only* the Scottish Supplement can amend or modify the Conditions.

Clause 3   In general, this clause is an agreement between the parties to settle by arbitration all disputes or differences in the Sub-contract. The remainder of the clause on agreement and disagreement as to the appointed Arbiter, and the time limit after written notice is self-explanatory.

Clause 3.1   This is another agreement that arbitration cannot commence without the written consent of the parties, until after determination of the Contractor under the Main Contract, Practical Completion of the Main Contract, or abandonment of the Main Contract Works. In other words, the ending of the Main Contract however it might come about.

Following the general part of the clause is a list of exceptions to the above.

It is of course, normal procedure to give a notice in writing to start arbitration proceedings at the time, not to wait until Practical Completion has been achieved, before a notice is issued.

Clause 3.2   Again, an agreement is made between the parties that if a similar dispute between the Employer and the Contractor is referred to an Arbiter, then the dispute between the Contractor and the Sub-contractor may be dealt with by the same Arbiter, if they both agree that he is suitably qualified.

Clause 3.3   This clause sets out in detail the powers of the Arbiter, and both parties here agree to abide by them.

Clause 3.4   The clause itself is self-explanatory, but it might be difficult to give a practical example.

Clause 3.5   Although this is an agreement to accept an Arbiter's award as final and binding, it affords the opportunity for either party to apply to the Court of Session for an opinion on a point of law. That opinion or judgement could be further appealed to the House of Lords. This can happen at any stage in the arbitration proceedings up until the Arbiter's award.

Clause 3.6   This gives the Arbiter the right to recover his fees and expenses.

Clause 4   The law of Scotland will operate and the Sub-contract is subject to same.

Clause 5   This clause can be deleted and initialled by both parties if they so desire. If it is left in, the clause causes the Contract to be registered in the Books of Council and Session. It is necessary to have the signed contract stamped with a 50p stamp at the Inland Revenue Office, Edinburgh.

The contract has to be signed on Page 3 in the presence of two witnesses to each signature and signed again on Page 5. As mentioned previously, legal advice on the signing of the contract is advisable.

Appendix – Scottish Supplement

As mentioned at the beginning, this allows the Sub-contract NSC/4 to be amended to accord with the Law of Scotland and Scottish practice.

Clause 1   This deletes Clauses 1.1 and 1.2 of the definitions and allows the Scottish version of the various forms to be substituted under Clause 1.3. Also, Clause 1.4 changes the English to Scottish wording.

Clause 19A and 19B   Either one or the other of these options must be deleted and be the same as that deleted in Schedule 2 of Tender NSC/1/Scot, Item 8. *See* the comments on Item 8 explanatory notes to Tender NSC/1.

Clauses 20A and 20B   The deletion made here will again depend upon the assertion made in Schedule 2 of Tender NSC/1/Scot, Item 7 and on Page 1 of Sub-contract NSC/4/Scot. *See* the comments on Item 7, explanatory notes to Tender NSC/1.

Clause 21   The two sub-clauses deleted are those dealing with the payment for off-site goods and materials.

The additional sub-clause is similar to the Supplement to the Main Contract, where off-site goods and materials become the subject of a Contract of Purchase and not at the Architect's discretion for inclusion in an Interim Certificate.

Clause 24   This is in lieu of Clause 6 in Schedule 2 of Tender NSC/1/Scot (dealing with an adjudicator and trustee stakeholder) which is not used.

In Scotland the Sub-contractor is entitled to appoint an adjudicator to be selected from a list maintained by the Scottish Building Contract Committee and monies shall be placed on deposit receipt in joint names of the Contractor and the Sub-contractor.

The sub-clauses also deal with the appointment of the Adjudicator, his selection of a bank, the deposition of monies and payment in accordance with the Arbiter's award.

Clause 29.2   In Scotland discretion is granted to the Contractor, if a provisional liquidator is appointed to control the Sub-contractor's business, either to determine or to wait for the Provisional Liquidator to complete his investigations. The Contractor must give the Provisional Liquidator seven days notice of determination. Bankruptcy is still a case for automatic determination.

Clause 31.3   This is an additional clause in Scotland in which the Sub-contractor agrees to allow the Contractor to assign his Sub-contract to the Employer, on the determination of the Contractor's employment, as long as it is not caused by bankruptcy, etc. The assignment shall be made within 14 days of determination and the Sub-contractor can make reasonable objection to any further assignment by the Employer.

This brings the Sub-contract into line with Clause 27.4.2.1 of the Main Contract Conditions.

Nomination procedure – Alternative Method

Again, there is a separate set of documents which are listed as the JCT and SBCC published documents at the beginning of this book. The documents have the same numbers but with the suffix 'Scot' added to distinguish them.

As in England and Wales, Scotland has no Tender document and Tender NSC/1/Scot must not be used for this purpose.

In Scotland the Standard Form of Nomination NSC/3a/Scot is available as a discretionary additional document for the Architect's use in this form of nomination.

Diagram 5(S) following shows the complete procedure from start to finish, all of which is covered in Clauses 35.11 and 35.12 of the Main Conditions.

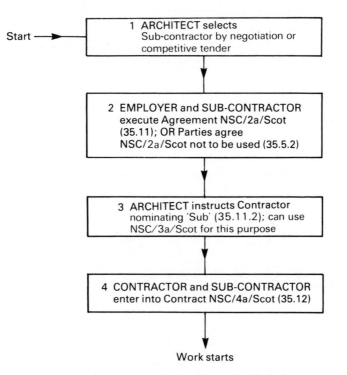

Start ⟶

1 ARCHITECT selects
Sub-contractor by negotiation or
competitive tender

2 EMPLOYER and SUB-CONTRACTOR
execute Agreement NSC/2a/Scot
(35.11); OR Parties agree
NSC/2a/Scot not to be used (35.5.2)

3 ARCHITECT instructs Contractor
nominating 'Sub' (35.11.2); can use
NSC/3a/Scot for this purpose

4 CONTRACTOR and SUB-CONTRACTOR
enter into Contract NSC/4a/Scot (35.12)

Work starts

Diagram 5(S): Nomination of a Sub-contractor using the Alternative Method

Diagram 6(S) shows the procedure if the proposed Sub-contractor fails to enter into a contract without due cause (Clause 35.23.3), while *Diagram 7(S)* shows the procedure if the Contractor makes a reasonable objection within seven days (Clause 35.4.3).

Immediately following the two diagrams are the forms, indicating where the various parties have to provide information and/or signatures, to effect a nomination, together with such additional explanatory notes as are necessary.

The timing of the giving of information and/or signatures is important and, with this in view, the various stage numbers to accord with *Diagram 5(S)* and name of party are also entered in the forms.

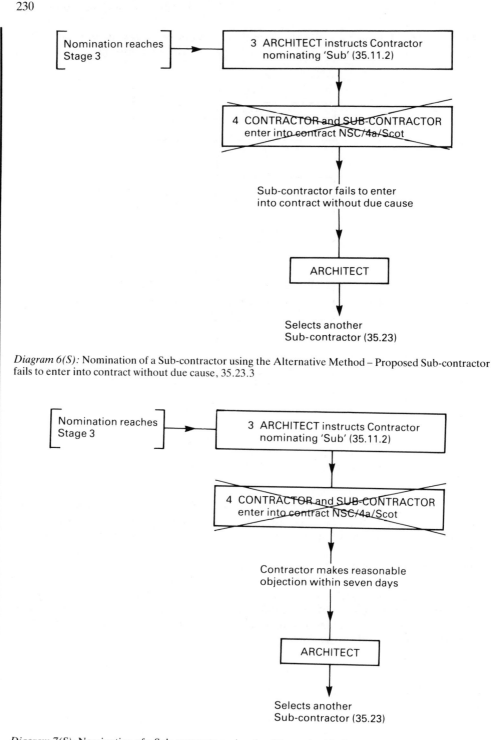

Diagram 6(S): Nomination of a Sub-contractor using the Alternative Method – Proposed Sub-contractor fails to enter into contract without due cause, 35.23.3

Diagram 7(S): Nomination of a Sub-contractor using the Alternative Method – Contractor makes reasonable objection within seven days, 35.4.3

N.B.

April 1981 Revision

As compared with the original the following alterations have been made.

1. Clause 1 – rewritten to comply with identical clause in English NSC/2a.

2. Clause 3 – correction of error:
   Reference to completion of Sub-Contract Works altered to Main Contract Works.

3. Clause 8 – addition to end of clause –
   excepting 35.24.6.

Agreement NSC/2a/Scot

(Revised April 1981)

SBCC

AGREEMENT

between

(Employer)

and

(Sub-Contractor)

This form is issued by the
Scottish Building Contract Committee of which the
constituent bodies are:

Royal Incorporation of Architects in Scotland

Scottish Building Employers Federation

Scottish Branch of the Royal Institution of
Chartered Surveyors

Convention of Scottish Local Authorities

Federation of Specialists and Sub-Contractors
(Scottish Board)

Committee of Associations of Specialist
Engineering Contractors (Scottish Branch)

Association of Consulting Engineers
(Scottish Group)

Confederation of British Industry

Association of Scottish Chambers of Commerce

Copyright of the S.B.C.C., 39 Castle Street, Edinburgh

April 1981

232

Agreement NSC/2a/Scot
(Revised April 1981)

Agreement between an Employer and a Sub-Contractor nominated under Clauses 35.11 and 35.12 of the Standard Form of Building Contract

Main Contract Works:

Location:

Sub-Contract Works:

AGREEMENT

between

ARCHITECT - STAGE 2

(hereinafter referred to as 'the Employer')

and

(hereinafter referred to as 'the Sub-Contractor')

_____

CONSIDERING THAT the Employer has entered into/intends to enter into a Contract for certain Works (hereinafter referred to as 'the Main Contract') for which the Contract Conditions are
(hereinafter referred to as 'the Main Contract Conditions'):

FURTHER CONSIDERING that the Employer has appointed

to be the Architect/Supervising Officer for the purposes of the Main Contract and this Agreement (who, and his validly appointed successors, is hereinafter referred to as 'the Architect/Supervising Officer'):

FURTHER CONSIDERING that the Sub-Contractor has tendered for certain Works (hereinafter referred to as 'the Sub-Contract Works') which tender dated_____ has been approved by the Architect/Supervising Officer on behalf of the Employer on the basis that Clauses 35.11 and 35.12 of the Main Contract Conditions shall apply:

Page 1

FURTHER CONSIDERING that the Architect/Supervising Officer has issued/intends to issue an instruction under the said Clause 35.11 nominating the Sub-Contractor to carry out and complete the Sub-Contract Works under the terms and conditions of Sub-Contract NSC/4a/Scot:

FURTHER CONSIDERING that the Employer and the Sub-Contractor have agreed to enter into this Agreement

T H E R E F O R E the Employer and the Sub-Contractor HAVE AGREED and DO HEREBY AGREE as follows:

1 The Sub-Contractor warrants that he has exercised and will exercise all reasonable skill and care in

 1.1 the design of the Sub-Contract Works in so far as the Sub-Contract Works have been or will be designed by him,

 1.2 the selection of materials and goods for the Sub-Contract Works in so far as such materials and goods have been or will be selected by him, and

 1.3 the satisfaction of any performance specification or requirement in so far as such performance specification or requirement is included or referred to in the description of the Sub-Contract Works included in or annexed to the Tender.

2 If after execution of this Agreement and before the issue by the Architect/ Supervising Officer of the nomination instruction under Clause 35.11 of the Main Contract Conditions the Architect/Supervising Officer instructs the Sub-Contractor in writing to proceed with the designing of or the proper ordering or fabrication of any materials or goods for the Sub-Contract Works, the Sub-Contractor shall forthwith comply with the instruction and

 2.1 the Employer shall pay the Sub-Contractor the amount of any expense reasonably and properly incurred by him in carrying out work in the designing of the Sub-Contract Works and upon such payment the Employer may use that design work for the purposes of the Sub-Contract Works but not further or otherwise;

 2.2 the Employer shall pay the Sub-Contractor for any materials or goods properly ordered or fabricated by the Sub-Contractor for the Sub-Contract Works and upon such payment any materials and goods shall become the property of the Employer;

 2.3 the Employer shall not be required to make any payment under Clauses 2.1 and 2.2 hereof after the issue of the nomination instruction under Clause 35.11 of the Main Contract Conditions except in respect of any design work properly carried out and/or materials or goods properly ordered or fabricated in compliance with an instruction under Clause 2 hereof but which are not used for the Sub-Contract Works by reason of some written decision of the Architect/Supervising Officer before the issue of the nomination instruction under Clause 35.11 of the Main Contract Conditions;

 2.4 if the Employer has made any payment under Clauses 2.1 and 2.2 hereof and the Sub-Contractor is subsequently nominated to execute the Sub-Contract Works by virtue of a nomination instruction issued under Clause 35.11 of the Main Contract Conditions the Sub-Contractor shall allow the Employer and the Main Contractor full credit for such payment in the discharge of the amount due in respect of the Sub-Contract Works.

3 The Sub-Contractor shall supply the Architect/Supervising Officer with such information (including drawings) in accordance with the agreed programme details or at such time as the Architect/Supervising Officer may reasonably require so that the Architect/Supervising Officer shall not be delayed in issuing necessary instructions or drawings under the Main Contract for which delay the Contractor might have a valid claim for an extension of time for the completion of the Main Contract Works under Clause 25.4.6 or a valid claim for direct loss and expense under Clause 26.2.1 of the Main Contract Conditions.

4 The Sub-Contractor shall perform his obligations under the Sub-Contract so that the Architect/Supervising Officer will not by reason of any default of the Sub-Contractor be required to issue an instruction to determine the employment of the Sub-Contractor under the Main Contract: Provided that any suspension by the Sub-Contractor of further execution of the Sub-Contract Works under Clause 21.8 of Sub-Contract NSC/4a shall not be regarded as coming within this clause.

5 The Sub-Contractor shall perform the Sub-Contract so that the Contractor will not become entitled to an extension of time for completion of the Main Contract Works.

6 The Employer agrees that the Sub-Contractor will not be liable under Clauses 3, 4 and 5 hereof until the Architect has issued his nomination instruction under Clause 35.11 of the Main Contract Conditions.

*Clause 35.13.1 requires the Architect/Supervising Officer when directing the Contractor as to the amount included in an Interim Certificate in respect of the value of the Nominated Sub-Contract Works issued under Clause 30 of the Main Contract Conditions forthwith to inform the Sub-Contractor in writing of that amount

†Clause 35.17 deals with final payment by the Employer for the Sub-Contract Works prior to the issue of the Final Certificate under the Main Contract Conditions

7 The Employer shall ensure that the Architect/Supervising Officer will operate the provisions of Clauses 35.13.1* and 35.17† to 35.19 of the Main Contract Conditions.

7.1 After final payment to the Sub-Contractor under Clause 35.17 of the Main Contract Conditions the Sub-Contractor shall rectify at his own cost (or if he fails to do so shall be liable to the Employer for the costs referred to in Clause 35.18 of the Main Contract Conditions) any omission, fault or defect in the Sub-Contract Works which the Sub-Contractor is bound in accordance with the Sub-Contract to rectify, after written notification of the same by the Architect/Supervising Officer at any time prior to the issue of the Final Certificate under the Main Contract.

7.2 After issue of the said Final Certificate the Sub-Contractor shall in addition to such other responsibilities if any, as he has under this Agreement, have the same responsibility to the Contractor and to the Employer for the Sub-Contract Works as the Contractor has to the Employer under the Main Contract up to the issue of the said Final Certificate.

8 Where the Architect/Supervising Officer has been required under Clause 35.24 of the Main Contract Conditions to make a further nomination in respect of the Sub-Contract Works, the Sub-Contractor shall indemnify the Employer for any direct loss and/or expense resulting therefrom, except where Clause 35.24.6 applies.

9 The Employer shall operate, and shall ensure that the Architect/Supervising Officer shall operate, the provisions in regard to the payment of the Sub-Contractor contained in Clause 35.13 of the Main Contract Conditions.

9.1 If, after paying any amount to the Sub-Contractor under Clause 35.13.5.3 of the Main Contract Conditions, the Employer produces reasonable proof that there was in existence at the time of such payment a petition or resolution to which Clause 35.13.5.4.4 of the Main Contract Conditions refers, the Sub-Contractor shall repay such amount, on demand.

*Clause 2.3 deals with specified supplies and restrictions etc. in the contracts of sale for such supplies

10 Where Clause 2.3* of Sub-Contract NSC/4a applies the Sub-Contractor shall forthwith give the Contractor details of any restriction, limitation or exclusion to which that clause refers as soon as such details are known to the Sub-Contractor.

11 In the event of any dispute or difference of opinion arising regarding any term of this Agreement or as to any matter or thing arising hereunder or in connection herewith the same shall be and is hereby referred to the decision of an Arbiter mutually agreed on, or failing such agreement within fourteen days after either party has given written notice to the other to concur in the appointment of an Arbiter, by an Arbiter appointed by the Sheriff of any Sheriffdom in which the Sub-Contract Works or any part thereof are situated.

11.1 The Arbiter shall without prejudice to the generality of his powers have power to make such orders and directions as may in his opinion be necessary to determine the rights of the parties and to open up review and revise any directions, requirements or notices given by either party to the other, and to determine all matters in dispute as if no such direction, requirement or notice had been given.

11.2 The Law of Scotland shall apply to all arbitrations under these presents, and the Award of the Arbiter shall be final and binding on the parties, subject always to the provisions of Section 3 of the Administration of Justice (Scotland) Act 1972.

11.3 The Arbiter shall be entitled to remuneration and reimbursement of his outlays.

236

12 This Agreement shall be construed and the rights of the Employer and the Sub-Contractor and all matters arising hereunder shall be determined according to the Law of Scotland: IN WITNESS WHEREOF these presents are executed

at   on the   day of   19

before these witnesses subscribing:

_____witness

_____address

_____occupation

EMPLOYER - STAGE 2

        _____Employer.

_____witness

_____address

_____occupation

_____witness

_____address

_____occupation

SUB-CONTRACTOR -STAGE 2

        _____Sub-Contractor.

_____witness

_____address

_____occupation

N.B.–This document is set out for execution by individuals or firms: Where Limited Companies or Local Authorities are involved amendment will be necessary and the appropriate officials should be consulted.

Explanatory notes

Agreement NSC/2a/Scot is a yellow-coloured document which gives the same information as Agreement NSC/2/Scot but in a slightly different format with the omission of the non-applicable clauses.

It must be understood that the Sub-contractor's Tender has to be approved by the Architect before this Agreement is entered into.

Again, the Architect must not sign the Agreement on behalf of the Employer since in itself it is a binding contract which both parties sign in the presence of two witnesses. The Employer must be one of the signatories.

The use of Agreement NSC/2a/Scot is optional when using the Alternative Method. just as it is in England and Wales. The choice as to whether or not to use it has already been fully explained. Remember that its inclusion in the Alternative Method will have been stated to the Contractor when a nomination was intended and to the Sub-contractor with his invitation to tender.

A comparison of NSC/2a/Scot with NSC/2/Scot will show differences in the number of and numbering of clause, NSC/2a/Scot being a shorter document. Explanatory notes both for NSC/2/Scot and NSC/2 cover all the clauses in this document and can therefore be read in reference to this Agreement, NSC/2a/Scot.

238

Standard Form for Nomination of Sub-Contractors for use in Scotland
(where Tender NSC/1/Scot has not been used)

To (Main Contractor)

ARCHITECT - STAGE 3
AT HIS DISCRETION

Main Contract Works

Sub-Contract Works

Page(s) number of Bills
or Specification

Name and address of
Nominated Sub-Contractor

Further to   (1) Any preliminary notice of nomination given

    and (2) The tender completed by the Sub-Contractor referred to above and approved by me/us on behalf of the Employer

I/we hereby nominate the Sub-Contractor referred to above under Clauses 35.11 and 35.12 of the Main Contract.

_____
Architect/Supervising Officer

Date_____

Issued to	Main Contractor	Q.S.	Nominated Sub-Contractor	Clerk of Works	Consulting Engineer	A/S.O. file
	☐	☐	☐	☐	☐	☐

Copyright SBCC, 39 Castle Street, Edinburgh   February 1980

Explanatory notes

Nomination NSC/3a/Scot (yellow-coloured) is an option for the Architect in issuing his nomination instruction. It can be used by the Architect if he considers it convenient to do so.

Notice the change in wording from NSC/3/Scot as regards preliminary notice of nomination. In NSC/3/Scot a preliminary notice has been given, whereas in NSC/3a/Scot there is no legal requirement for a preliminary notice to have been issued. Also, in both versions the tender has to be approved by the Architect, although in regard to NSC/3/Scot it is Tender NSC/1/Scot which is referred to, and in NSC/3a/Scot it is the Sub-contractor's own tender.

The explanatory notes for Nomination NSC/3 in England and Wales are also relevant here.

240

N.B.

April 1981 Revision

As compared with the original the following alterations
have been made.

Building Contract
  Clause 3.1.1 – Delete 'certificate:'
  substitute 'payment.'
  Clause 3.1.4 – Add reference to 4.3

Appendix No. I
  15 – Price for Sub-Contract Works
    Add amendments to 15.1 and 15.2
  21 – Payment of Sub-Contractor
    Add amendment to 21.3.2.1

Appendix No. II
  Part 2 Section A
  Reference to Master Programme included
  Part 5
  Reference to Dayworks percentages included
  Part 10
  Reference to Basic Transport charge list included.

Scottish Building Sub-Contract NSC/4a/Scot

(Revised April 1981)

SBCC

BUILDING SUB-CONTRACT

between

and

with

SCOTTISH SUPPLEMENT 1980

to

The Conditions of the Standard Form of Sub-Contract
for Sub-Contractors nominated under Clauses 35.11/35.12 of
the Standard Form of Building Contract
1980 Edition

The constituent bodies of the
Scottish Building Contract Committee are:

Royal Incorporation of Architects in Scotland
Scottish Building Employers Federation
Scottish Branch of the Royal Institution of Chartered Surveyors
Convention of Scottish Local Authorities
Federation of Specialists and Sub-Contractors
  (Scottish Board)
Committee of Associations of Specialist Engineering
  Contractors (Scottish Branch)
Association of Consulting Engineers (Scottish Group)
Confederation of British Industry
Association of Scottish Chambers of Commerce

Copyright of the S.B.C.C., 39 Castle Street, Edinburgh

April 1981

Scottish Building Sub-Contract NSC/4a/Scot
(Revised April 1981)

For use in Scotland for Sub-Contractors nominated under Clauses 35.11 and 35.12 of the Standard Form of Building Contract

Main Contract Works:

Location:

Sub-Contract Works:

BUILDING SUB-CONTRACT

between

CONTRACTOR OR ARCHITECT STAGE 4

(hereinafter referred to as 'the Contractor')

and

(hereinafter referred to as 'the Sub-Contractor')

CONSIDERING THAT the Sub-Contractor has submitted a Tender for the Sub-Contract Works described in numbered documents (the said Sub-Contract Works and numbered documents being referred to in Part 1 of Appendix No. 1 thereto) which are to be executed as part of the Main Contract Works referred to in Part 2 of the said Appendix:

*Delete if NSC/2a/Scot is not required

FURTHER CONSIDERING that the Architect/Supervising Officer in the Main Contract has selected and approved the Sub-Contractor to carry out the Sub-Contract Works and *the Employer in the Main Contract and the Sub-Contractor have entered into the Agreement NSC/2a/Scot:

FURTHER CONSIDERING that the Architect/Supervising Officer has issued an instruction (with a copy to the Sub-Contractor) dated_____ under Clause 35.11 of the Main Contract nominating the Sub-Contractor to carry out the Sub-Contract Works and the Contractor by virtue of Clause 35.12 of the Main Contract Conditions is required to enter into these presents with the Sub-Contractor within 14 days of the date of the said nomination instruction:

FURTHER CONSIDERING that at the date of this Building Sub-Contract

*Delete as applicable

(a) the Sub-Contractor is*/is not the user of a current Sub-Contractor's Tax Certificate under the provisions of the Finance (No. 2) Act 1975 in one of the forms specified in Regulation 15 of the Income Tax (Sub-Contractors in the Construction Industry) Regulations 1975 and the Schedule thereto:

(b) the Contractor is*/is not the user of a current Sub-Contractor's Tax Certificate under the said Act and Regulations; and

(c) the Employer under the Main Contract is*/is not 'a Contractor' within the meaning of the said Act and Regulations:

THEREFORE the Contractor and the Sub-Contractor HAVE AGREED and DO HEREBY AGREE

*Complete as required. The alternative is for use when the Sub-Contract Works are to be completely remeasured and valued

1   The Sub-Contractor will carry out and complete the Sub-Contract Works for
     *the sum of

     (£   ) (hereinafter referred to as 'the Sub-Contract Sum') or such other
     sum as shall become payable in accordance with this Building Sub-Contract.

     OR

     *the sum of

     (£   ) (hereinafter referred to as 'the Tender Sum') as finally
     ascertained in accordance with this Building Sub-Contract.

2

2   The Sub-Contract Works shall be carried out and completed in accordance with, and the rights and duties of the Contractor and Sub-Contractor shall be regulated by Conditions of the Standard Form of Sub-Contract for Sub-Contractors nominated under Clauses 35.11 and 35.12 of the Standard Form of Building Contract (known as Sub-Contract NSC/4a) issued by the Joint Contracts Tribunal which are held to be incorporated in and form part of this Sub-Contract subject only to the amendments and modifications thereto contained in the Scottish Supplement forming Appendix No. I hereto and to the terms and requirements of Appendix No. II hereto.

3   In the event of any dispute or difference between the Contractor and the Sub-Contractor arising during the progress of the Sub-Contract Works or after completion or abandonment thereof in regard to any matter or thing whatsoever arising out of this Sub-Contract or in connection herewith, then such dispute or difference shall be and is hereby referred to the arbitration of such person as the parties may agree to appoint as Arbiter or failing agreement within 14 days after either party has given to the other written notice to concur in the appointment of an Arbiter as may be appointed by the Sheriff of any Sheriffdom in which the Sub-Contract Works or any part thereof are situated: Arbitration proceedings shall be deemed to have been instituted on the date on which the said written notice has been given.

3.1   No arbitration shall commence without the written consent of the parties until after determination or alleged determination of the Contractor's employment under the Main Contract or until after Practical Completion or alleged Practical Completion of the Main Contract Works or abandonment of the Main Contract Works unless it relates to

3.1.1   whether or not a payment has been improperly withheld or is not in accordance with the said Conditions

3.1.2   whether Practical Completion of the Sub-Contract Works has taken place in accordance with Clause 14.2

3.1.3   a claim by the Contractor or counter-claim by the Sub-Contractor under Clause 24

3.1.4   a dispute under Clauses 4.3, 11.2 or 11.3.

3.2   If the dispute or difference is substantially the same as or is connected with a dispute or difference between the Contractor and the Employer under the Main Contract the Contractor and Sub-Contractor hereby agree that such dispute or difference shall be referred to an Arbiter appointed or to be appointed to determine the related dispute or difference; Provided that either party may require the appointment of a different Arbiter if he reasonably considers the Arbiter appointed in the related dispute is not suitably qualified to determine the dispute or difference under this Sub-Contract.

3.3   The Arbiter shall have power to

3.3.1   direct such measurements and/or valuations as may in his opinion be desirable in order to determine the rights of the parties

3.3.2   ascertain and award any sum which ought to have been referred to or included in any certificate

3.3.3   open up review and revise any certificate, opinion, decision, requirement or notice

3.3.4   determine all matters in dispute which shall be submitted to him in the same manner as if no such certificate, opinion, decision, requirement or notice had been given

3.3.5   award compensation or damages and expenses to or against any of the parties to the arbitration.

3.4   In any arbitration under this clause a decision of the Architect which is final and binding on the Contractor shall also be final and binding between and upon the Contractor and the Sub-Contractor.

3.5   The Law of Scotland shall apply to all arbitrations in terms of this clause and the award of the Arbiter shall be final and binding on the parties subject to the provisions of Section 3 of the Administration of Justice (Scotland) Act 1972.

3.6   The Arbiter shall be entitled to remuneration and reimbursement of his outlays.

3

4 Both parties agree that this Sub-Contract shall be regarded as a Scottish Contract and shall be construed and the rights of parties and all matters arising hereunder determined in all respects according to the Law of Scotland.

5 Both parties consent to registration hereof for preservation and execution:

IN WITNESS WHEREOF these presents are executed at

on the

day of   19   before these witnesses subscribing.

_____witness

_____address

_____occupation   CONTRACTOR
                                     -STAGE 4
                                     _____ Contractor.

_____witness

_____address

_____occupation

_____witness

_____address

_____occupation   SUB-CONTRACTOR
                                     -STAGE 4
                                     _____ Sub-Contractor.

_____witness

_____address

_____occupation

N.B. – This document is set out as for execution by individuals or firms: Where Limited Companies are involved amendment will be necessary and the appropriate officials should be consulted.

Both parties sign here and on pages 5 and 10.

4

SCOTTISH SUPPLEMENT

(The following are the amendments and modifications to the
Conditions of Standard Form of Sub-Contract NSC/4a.
The numbers refer to clauses in NSC/4a)

1   Interpretation, Definitions, etc.

1.1 and 1.2 shall be deleted.

1.3   The meanings given to the under noted words and phrases shall be deleted and the
following substituted:

Agreement NSC/2	Agreement NSC/2a/Scot
Appendix	The completed Appendix II to the Building Sub-Contract NSC/4a/Scot
Arbitrator	Arbiter
Article or Articles of Sub-Contract Agreement	The foregoing Building Sub-Contract
Sub-Contract NSC/4a	Building Sub-Contract NSC/4A/Scot

The following clause shall be added:

1.4   Additional definition:

Real or personal	Heritable or moveable.

15   Price for Sub-Contract Works

15.1   The words 'Where Article 2.1 applies' shall be deleted and the words 'Where the
Sub-Contractor has quoted a Sub-Contract Sum' substituted.

15.2   The words 'Where Article 2.2 applies' shall be deleted and the words 'Where the
Sub-Contractor has quoted a Tender Sum' substituted.

19A and 19B – Value Added Tax

*Delete as applicable

*19A shall apply.

*19B shall apply.

*N.B. (1) Clause 19B applies only when the Main Contractor under the VAT (General) Regulations
1972 (Regulation 8(3)) has with the approval of the Sub-Contractor been allowed to prepare
tax documents in substitution for an authenticated receipt issued by the Sub-Contractor under
Regulation 21(2).*

(2) Where Clause 19B applies Clause 19A must be deleted.

20A and 20B – Finance (No. 2) Act 1975 – Tax Deduction Scheme

*Delete as applicable

*20A shall apply.

*20B shall apply.

*N.B. Clause 20A applies when the Sub-Contractor has a current Sub-Contractor's Tax
Certificate.*

*Clause 20B applies when the Sub-Contractor has no such Certificate. See narrative of the
foregoing Building Sub-Contract.*

20A.3.2/20B.3 shall be deleted and the following substituted:
The Contractor shall immediately inform the Sub-Contractor of any change in his own
status or that of the Employer as stated in the foregoing Building Sub-Contract.

20A.8/20B.6   There shall be deleted 'the provisions of Article 3 shall apply to' and added at the
end 'shall be referred to an Arbiter appointed in accordance with the foregoing Building
Sub-Contract.'

21   Payment of Sub-Contractor

21.2.3 shall be deleted.

21.3.2.1 line 4 The words 'clause 4' shall be deleted, and the words 'clause 7' substituted.

21.4.1.3 shall be deleted.

The following clause shall be added:

21.4.4   If the Architect/Supervising Officer is of the opinion that it is expedient to do so, the
Sub-Contractor may with the consent of the Contractor, which consent shall not be
unreasonably withheld, enter into a Contract for the purchase of materials and/or goods
by the Employer in the Main Contract prior to their delivery to the site, and upon such
Contract being entered into the purchase of the said off-site materials and/or goods
shall be excluded altogether from this Sub-Contract and the Sub-Contract Sum or the
Tender Sum shall be adjusted accordingly: And it is specifically declared and provided
that payment by the Employer to the Sub-Contractor for any of the said materials
and/or goods shall in no way affect any cash discounts or other emoluments to which the
Contractor may be entitled.

5

24 Contractors claims not agreed by the Sub-Contractor – appointment of Adjudicator

24.1.2 shall be deleted and the following substituted:

Subject to the provisions of Clauses 21.3.23 and 24 hereof the Sub-Contractor shall be entitled to appoint an Adjudicator to be selected from a list maintained by the Scottish Building Contract Committee to decide those matters referable to the Adjudicator under the provisions of Clause 24. In the event of the first person approached declining to act the Sub-Contractor shall approach another person on the said list and so on until the appointment is made: Provided that no Adjudicator shall be appointed who has any interest in this Sub-Contract or the Main Contract of which this Sub-Contract is a part or in other Contracts or Sub-Contracts in which the Contractor or the Sub-Contractor is engaged, unless the Contractor, Sub-Contractor and Adjudicator so mutually agree in writing.

24.3.1.2 shall be deleted and the following substituted:

shall be placed on Deposit Receipt in joint names of the Contractor and Sub-Contractor with such Bank as the Adjudicator shall direct pending the result of the arbitration.

24.4.1 shall be deleted and the following substituted:

Where any decision of the Adjudicator notified under Clause 24.3.3 requires the Contractor to place an amount on joint Deposit Receipt as aforesaid the Contractor shall forthwith do so and shall send a duplicate of the said Deposit Receipt to the Sub-Contractor immediately on issue: Provided that the Contractor shall not be obliged to deposit a sum greater than the amount due from the Contractor under Clause 21.3 hereof in respect of which the Contractor has exercised the right of set-off referred to in Clause 23.2 hereof.

24.5 shall be deleted and the following substituted:

Any amount placed on joint Deposit Receipt as aforesaid shall remain on deposit until the Arbiter appointed pursuant to the aforesaid notice of arbitration directs that it should be uplifted or the Contractor and Sub-Contractor mutually so agree. In either case the said Deposit Receipt shall be endorsed by the Contractor and the Sub-Contractor and the sum on Deposit Receipt (with interest if appropriate) uplifted and paid to the Contractor or the Sub-Contractor in accordance with the Arbiter's award or as the Contractor and Sub-Contractor may have mutually agreed as the case may be.

24.8 There shall be deleted 'and where relevant, for the charges of the Trustee-Stakeholder or any part thereof.'

29 Determination of the employment of the Sub-Contractor by the Contractor

29.2 shall be deleted and the following substituted:

In the event of a Provisional Liquidator being appointed to control the affairs of the Sub-Contractor the Contractor may determine the employment of the Sub-Contractor under this Contract by giving him 7 days written notice sent by Registered Post or Recorded Delivery of such determination. In the event of the Sub-Contractor becoming bankrupt or making a composition or arrangement with his creditors or having his estate sequestrated or being rendered notour bankrupt or entering into a Trust Deed with his creditors or having a winding up order made or (except for the purpose of reconstruction) a resolution for voluntary winding up passed or a Receiver or Manager of his business or undertaking duly appointed or possession taken by or on behalf of the holders of any Debenture secured by a Floating Charge then without prejudice to any other rights or remedies which the Contractor may have the employment of the Sub-Contractor under this Sub-Contract shall be automatically determined.

29.3.3 The words 'any such property of the Sub-Contractor' shall be deleted and the words 'any such property so far as belonging to the Sub-Contractor' substituted.

31 Determination of the Main Contractor's employment under the Main Contract

The following clause shall be added:

31.3 The Sub-Contractor shall recognise an assignation by the Contractor in favour of the Employer in terms of Clause 27.4.2.1 of the Main Contract Conditions.

STAGE 4 _____Contractor.   STAGE 4 _____Sub-Contractor.

6

PART 1

Particulars of the Sub-Contract Works

Numbered Documents annexed to Sub-Contract NSC/4a/Scot
(to be listed here)

CONTRACTOR
- STAGE 4

PART 2

Section A   Main Contract Works:
(insert same description
as in the Main Contract)

Employer:

*Delete as applicable

Form of Main Contract Conditions: Standard Form of Building Contract 1980 Edition
Local Authorities Edition*/Private Edition
With Quantities*/Without Quantities*/With
Approximate Quantities (revised   ) all as
amended and modified by the
Scottish Supplement*/Scottish Supplement for
Sectional Completion

Inspection of Main Contract: The unpriced Bills of Quantities*/Bills of
Approximate Quantities*/Contract Specification
(incorporating the general conditions and
Preliminaries of the Main Contract) and the
Contract Drawings may be inspected at:

Main Contract Conditions: Works insurance: Clause 22A*/22B*/22C.
alternative provisions   Master programme: Clause 5.3.1.2 deleted*/not
deleted
Insurance: Clause 21.2.1.
Provisional sum included*/not included.

Main Contract Conditions:
any changes from printed Form
identified above:

7

Section B Main Contract: Appendix No. II and entries therein:

This Section will require to be amended when Sectional Completion applies to the Main Contract.

*Delete as applicable

	Clause	
Statutory tax deduction scheme – Finance (No. 2) Act 1975	31	Employer at Date of Tender is a 'Contractor'*/is not a 'Contractor' for the purposes of the Act and the Regulations.
Defects Liability Period (if none other stated is 6 months from the day named in the Certificate of Practical Completion of the Works)	17.2	
Insurance cover for any one occurrence or series of occurrences arising out of one event	21.1.1	£
Percentage to cover professional fees	22A	
Date of Possession	23.1	
Date for Completion	1.3	
Liquidate and Ascertained Damages	24.2	at the rate of £ _____ per _____
Period of delay: by reason of loss or damage caused by any one of the Clause 22 Perils	28.1.3.2	
for any other reason	28.1.3.1 and 28.1.3.3 to .7	
Period of Interim Certificates (if none stated is one month)	30.1.3	
Retention Percentage (if less than five per cent)	30.4.1.1	
Period of Final Measurement and Valuation (if none stated is 6 months from the day named in the Certificate of Practical Completion of the Works)	30.6.1.2	
Period for issue of Final Certificate (if none stated is 3 months)	30.8	
Work reserved for Nominated Sub-Contractors for which the Contractor desires to tender	35.2	

CONTRACTOR – STAGE 4

8

		Clause

*Delete as applicable

Fluctuations (if alternative desired is not shown Clause 38 shall apply)   37   Clause 38*/Clause 39*/Clause 40

Percentage addition   38.7 or 39.8 _____

Formula Rules   40.1.1.1

    Rule 3   Base Month _____19___

    Rule 3   Non-Adjustable Element _____%
                   (Local Authority with quantities only)

    Rules 10 and 30(i)   Part I*/Part II of Section 2 of the Formula Rules is to apply.

*This information might conveniently be given by attaching a copy of the Preliminaries Bill of the Main Contract

Section C   *Obligations or restrictions imposed by the Employer not covered by Main Contract Conditions (e.g. in Preliminaries in the Contract Bills)

CONTRACTOR - STAGE 4

PART 3

Insurance cover for any one occurrence or series of occurrences arising out of one event   7.1

£ _____

*Complete as far as relevant

PART 4*   11.1

The period/periods when Sub-Contract Works can be carried out on site:

to be between_____ and _____

Period required by Architect to approve drawings after submission   _____

Periods required:

(1) for submission of all further sub-contractor's drawings etc. (co-ordination, installation, shop or builder's work or other as appropriate*)   _____

*Not including period required for Architect's approval

(2) for execution of Sub-Contract Works: off-site   _____

                                   on-site   _____

(3) Notice required to commence work on site   _____

PART 5

Daywork percentages   16.3.4/17.4.3

Applicable Definition	Labour	Materials	Plant
RICS/NFBTE	____%	____%	____%
RICS/ECA/ECA of S	____%	____%	____%
RICS/HVCA	____%	____%	____%

Retention Percentage   21.5   _____%
(To be the same as that set out or referred to in Part 2 – Main Contract Appendix II and entries therein, Main Contract Conditions reference Clause 30.4.1.1).

PART 6

This part, which relates to the appointment of an Adjudicator and Trustee Stakeholder, is not applicable in Scotland but the number has been retained to keep the same numerical sequence as in England.

9

PART 7

Attendance (other than general attendance referred to in Clause 27.1.1).
(Details to be set out on a separate sheet(s) and attached to this appendix).

Clause

(1) Special scaffolding or scaffolding 27
additional to the Contractor's
standing scaffolding.

(2) The provision of temporary access
roads and hardstandings in
connection with structural steelwork, precast concrete components,
piling, heavy items of plant and
and the like.

(3) Unloading, distributing, hoisting
and placing in position giving in
the case of significant items the
weight and/or size (to be at the
risk of the Sub-Contractor).

(4) The provision of covered storage
and accommodation including
lighting and power thereto.

(5) Power supplies giving the
maximum load.

(6) Maintenance of specific temperature
or humidity levels.

(7) Any other attendance not included
under (1) to (6) or as general
attendance under Clause 27.1.1.

CONTRACTOR
-STAGE 4

PART 8

*Delete as required

Fluctuations   34   Clause 35*/Clause 36*/Clause 37.

PART 9

List of materials, goods, electricity   35   (To be set out on a separate sheet(s) and
including*/not including fuels   35.2.1   attached to this Appendix).

Date of Tender   35.6.1   _____

Percentage   35.7   _____ %

PART 10

Basic transport charge list   36.1.5   (To be set out on a separate sheet(s) and
                                                 attached to this Appendix)

Materials, goods, electricity   36.3.1   (To be set out on a separate sheet(s) and
including*/not including fuels   attached to this Appendix)
List of basic prices

Date of Tender   36.7   _____

Percentages   36.8   _____

PART 11   37

The JCT Nominated Sub-Contract   37.1   are those
Formula Rules   Dated_____19____
                                                 Part I*/Part III of these Rules applies.

*Not to exceed 10%

*Non-Adjustable Element   37.3.3   _____ %
(Local Authority with quantities only).
(See also Clause 37.3.4).

List of market prices   37.4   (To be set out on a separate sheet(s) and
                                                 attached to this Appendix).

10

JCT Nominated Sub-Contract Formula Rules

Rule 3   (Definition of Balance of Adjustable Work)

Any measured work not allocated to a Work Category.

CONTRACTOR
- STAGE 4

Base Month _____

Date of Tender_____

*Delete as applicable

Rule 8   Method of dealing with 'Fix-only' work. Rule 8(i)*/8(ii)*/8(iii) shall apply.

Rule 11a   Part I only: the Work Categories applicable to the Sub-Contract Works.

If both specialist engineering formulae apply to the Sub-Contract the percentages for use with each formulae should be inserted and clearly identified. The weightings to sprinkler installations may be inserted where different weightings are required

Rule 43   Part III only: Weightings of labour and materials – Electrical installations or Heating, Ventilating and Air Conditioning Installations.

	Labour	Materials
Electrical	_____%	_____%
Heating, Ventilating and Air Conditioning	_____%	_____%
Sprinklers	_____%	_____%

*Delete as applicable

Rule 61a   Adjustment shall be effected

*upon completion of manufacture of all fabricated components

*upon delivery to site of all fabricated components.

Rule 64   Part III only: Structural Steelwork Installations

(i)   Average price per tonne of steel delivered to fabricator's works   £_____

(ii)   Average price per tonne for erection of steelwork   £_____

Rule 70a   Catering Equipment Installations:

Apportionment of the value of each item between

(i)   Materials and shop fabrication   £_____

(ii)   Supply of factor items   £_____

(iii)   Site installations   £_____

STAGE 4 _____Contractor.   *STAGE 4* _____Sub-Contractor.

Explanatory notes

Both England-and-Wales and Scotland use the same set of Sub-contract Conditions, namely NSC/4a for this the Alternative Method.

In Scotland it is a Building Sub-contract that is completed by the Contractor and the Nominated Sub-contractor, not Articles of Agreement as in England and Wales. It is obligatory on the Contractor to conclude this contract within 14 days of nomination by the Architect. Again, differing from South of the Border, this yellow form is detached from the Sub-contract Conditions, which themselves are incorporated only by reference. The form also contains two appendices, Appendix I being the Scottish Supplement (as in NSC/4/Scot) and Appendix II being a transference of the Main Contract Appendix II details, together with agreed details as similar to Tender NSC/1/Scot in the Basic Method.

Appendix II is necessary because there is no standard form of tender for the Alternative Method. Accordingly, in the case of Sub-contract NSC/4a, which is common to both North and South of the Border, details to be agreed are incorporated by transfer to Appendix II of the Scottish Building Sub-Contract NSC/4a/Scot. In the case of Sub-contract NSC/4, which is also common to North and South of the Border, details to be agreed are stated in Tender NSC/1/Scot.

Although both Contractor and Nominated Sub-contractor have copies of the completed Scottish Building Sub-contract NSC/4a/Scot, it is still necessary to have a separate copy of Sub-contract Conditions NSC/4a for reference.

Scottish Building Sub-contract

The first paragraph states the Sub-contractor to have submitted a tender which is not in any standard form as already mentioned. It further states that all documents forming that Tender will be numbered for identification for later reference in Part 1 of Appendix II. Remember that the proposed sub-contractor does not have the benefit of Tender NSC/1/Scot where all the requirements of the Sub-contract are set out. It is therefore important that the tenderer has full knowledge of the Main Contract Conditions and know how Sub-contract NSC/4a/Scot will have to be completed, especially the Appendices. Notice also that the Tender documents will give a full description of the Sub-contract works. Reference can be made here to the explanatory notes for Tender NSC/1 in regard to the importance of this description and its ramifications.

The second paragraph requires the option to be stated as to whether or not the Agreement NSC/2a/Scot will be executed. If not, then the deletion as marked must be made. The Tenderer will have already been informed in his invitation to tender, whether or not the Agreement is to be executed.

The third paragraph states the date of the Instruction of Nomination and that the Contractor must have entered into a contract with the Nominated Sub-contractor within 14 days (Clause 35.12).

The fourth paragraph, similar to the fifth in NSC/4/Scot, states by deletion whether or not each party is a contractor in the terms of the Finance (No 2) Act 1975.

Clause 1   Here the Sub-contract Sum or the Tender Sum will be filled in. Be very careful that the correct sum is used and that the amount is exactly right.

Clause 2   This is where the reference to Sub-contract Conditions NSC/4a is made, and therefore incorporates it into Sub-contract NSC/4a/Scot. Note that *only* the Scottish Supplement can amend or modify the Conditions.

Clause 3   The comments on all these clauses for Sub-contract NSC/4/Scot are
to 5   relevant here, as the clauses are the same.

The contract has to be signed on Page 3 and on Pages 5 and 10 in the presence of two witnesses to each signature. As before, legal advice on the signing of the Contract is advisible.

Appendix I – Scottish Supplement

As mentioned before, this allows the Sub-contract NSC/4a to be amended to accord with the law of Scotland and Scottish practice.

Clause 1   This deletes Clauses 1.1 and 1.2 of the definitions and allows the Scottish version of the various forms to be substituted under Clause 1.3. Also Clause 1.4 changes the English to the Scottish wording.

Clauses 19A   Either one or the other of these options must be deleted. See the
and 19B   comments on Item 8 explanatory notes to Tender NSC/1.

Clauses 20A   The deletion here will depend on the assertion made on Page 1 of
and 20B   Sub-contract NSC/4a/Scot. See the comments on Item 7 explanatory notes to Tender NSC/1.

Clause 21   This is the same clause as in Sub-contract NSC/4/Scot, and the comments on that document are relevant here.

Clause 24   Again, this clause is explained in the explanatory notes to Sub-contract NSC/4/Scot, with the exception of the first paragraph, as there is no tender form in this form of nomination.

Clauses 29   These two clauses are the same as in Sub-contract NSC/4/Scot and are
and 31   explained in the explanatory notes thereto.

Appendix II

Part 1   The description should be in detail. The trade or specialism stated, any design element, any selection of materials and goods, any satisfaction as to a performance specification, or any other description necessary. All documents are numbered for identification, which means not just drawings but everything else making up the tender, such as specifications, Bills of Quantities, etc.

Part 2   This is a copy of the provisions of the Main Contract, all of which the tenderer should have been informed of at tendering stage. An incorrect transference of information here will stand, as this is the binding Sub-contract document.

Section A   The Conditions which govern the Main Contract are entered here. The information should be exact, with or without quantities, approximate quantities, sectional completion or whichever edition has been used, including any amendments.

The address where the Main Contract documents can be inspected is stated – ideally all documents should be kept at the same address.

The insurance options are stated, and two should be deleted to accord with the Main Contract.

Any other matters which have made a modification, amendment or supplementary clause to the Conditions, must be noted here.

Section B   This is a copy of the Main Contract Appendix II, and all the details from the Main Contract will be entered here with great care.

Section C   As the sidenote suggests, a blank copy of the Preliminaries Bill might well be attached. This would identify any Employer-imposed restrictions contained in the Main Contract, which would also then be imposed on this Sub-contract.

The Sub-contractor can make reasonable objection to variations on these restrictions, as stated in Sub-contract NSC/4a, Clause 4.3.

Part 3   This is the Sub-contractor's proportion of the insurance cover required by the Main Contract and is agreed with the Contractor. The entry requires the minimum amount for insurance cover to be inserted, as agreed, to cover any one occurrence or series of occurrences arising out of one event.

Part 4   As in Tender NSC/1/Scot of the Basic Method, these periods have to be agreed between the Contractor and the Sub-contractor. The first three items have to include time for approval and of course, if there is no design work involved, the necessary items deleted. It is probably more efficient if start and finish dates can also be inserted.

Remember that these dates constitute a base from which variations, extensions of time and direct loss and/or expense will operate, when the Sub-contract NSC/4a/Scot is executed.

Part 5   Not only is the retention percentage in the Main Contract stated here, but also the daywork percentage additions as required by Clauses 16.3.4.1 and 17.4.3.1 of the Sub-contract NSC/4a.

Part 6   This does not apply in Scotland but keeps the same numbering in sequence.

Part 7   The Sub-contractor has, without charge, general attendance from the Contractor, all as set out in Clause 27.1.1 of Sub-contract NSC/4a.

This Part lists in detail all the special attendance that the Sub-contractor will require. Note that separate sheets may be attached if the space is insufficient.

For detailed notes on Special Attendance see the explanatory notes to Tender NSC/1.

Part 8   Only one clause will apply, the other two being struck out. At tender stage one of the alternatives will have been selected. Note that if no selection has been made then Clause 35 applies.

Part 9   This deals only with the Clause 35 alternative, and the Architect's decision as to whether or not fuels is to be allowed will have been given to the Sub-contractor at time of tendering. The Sub-contractor lists the types or names of materials on which he wishes fluctuations to apply and he will be restricted to his stated list.

The date of tender and percentage will also have been given to the Sub-contractor by the Architect, at time of tendering.

Part 10   This deals only with the Clause 36 alternative and, when tendering, the Sub-contractor will have provided his list of basic prices to which he wishes fluctuations to apply. Again, the decision as to fuels will have been intimated to him. If the Sub-contractor follows the layout in Tender NSC/1 of four separate columns, this would be ideal. He is restricted to the lists of materials and prices that he gives in the applicable columns, for the award of fluctuations.

Part 11   This deals only with the Clause 37 alternative which is the Formula Adjustment Method.

With the following exceptions the explanatory notes to Schedule 1 Appendix B of Tender NSC/1 are relevant to this section.

The date of tender must be ten days before the date of the tender from the Sub-contractor. Whether it is a preliminary tender or a formal tender, it must be the tender from which the Contract Sum or Tender Sum was prepared.

The base month is one calendar month before the month in which the Tender is due to be returned, and that Tender is the one from which the Contract Sum or Tender Sum was prepared.

Note that whichever fluctuation is selected for use, the two alternatives will be struck out.

Chapter 13

Nominated Suppliers

England and Wales (Scotland, p. 268)

Clause 36 – Nominated Suppliers

The definition of a 'nominated supplier' is: a supplier to the Contractor who has been nominated by the Architect to supply materials and goods which are to be fixed by the Contractor.

Note that if a measured item in the Contract Bills specifies a sole supplier, this does not mean, nor is it intended to mean, that the supplier will be nominated. The Contractor could make objection to such a sole supplier but should do so before his tender is returned. If the Contractor feels that he will be obliged to accept any terms from that sole supplier which are abnormal to his usual trading procedures, he can make a case to the Architect for the supply of these goods and materials becoming a Prime Cost Sum for nomination. The Contractor can base his case upon the provisions of Clause 36.4, if any of those provisions listed are absent from the terms of sale, from the sole supplier.

Following is a tabular layout showing the methods of obtaining a nomination of a supplier by the Architect, with the relevant clause numbers.

Clause	Source	Nomination
36.1.1.1	Prime Cost Sum in the Contract Bills (a) If supplier named in Prime Cost Sum (b) Not named in Prime Cost Sum	Instruction under Clause 36.2 Named in Instruction under Clause 36.2
36.1.12	Provisional Sum in Contract Bills	Instruction with regard to expenditure of such sum, raising a Prime Cost Sum with the supplier named OR Instruction under Clause 36.2
36.1.13	Provisional Sum in Contract Bills, only one supplier	Instruction with regard to expenditure of such sum, raising a Prime Cost Sum in the Instruction and sole supplier is deemed to be nominated
36.1.1.4	None	Issue or sanction of variation which specifies a sole supplier, the supply made a Prime Cost Sum in the Instruction and sole supplier is deemed to be nominated

36.3   The adjustment of the Contract Sum under Clause 30.6.2.8 will be an amount which, where applicable, includes:

   (a) The total amount of the materials and goods less any discount, other than the 5 per cent cash discount.
   (b) Any tax (except VAT) or duty not otherwise recoverable under this Contract.
   (c) The net cost of packing, carriage and delivery, after allowance of any credit of packaging return to the Supplier.
   (d) The amount of any fluctuations paid to or allowed by the Supplier, less any discount other than for cash.

   36.3.2 This clause gives the Contractor the opportunity for reimbursement of any expense caused to him by his obtaining the materials and goods from the Nominated Supplier. The provisos are that he could not recover anywhere else in the Contract and that the Architect be in agreement. In this case, obviously, the Contractor would make a written claim.

36.4   Unless the Architect and the Contractor have a previous agreement, the Architect will nominate only a supplier who will enter into a contract of sale with the Contractor. However, by giving the Architect and the Contractor an alternative to make some other agreement, the possibility of the Contractor rejecting the nomination is avoided.

   This clause further lays down in sub-clauses the terms which have to be included in the contract of sale, namely:

   36.4.1 The materials and goods to be of the specified standard or to the Architect's satisfaction.

   36.4.2 The Nominated Supplier to make good any defects up till the end of the Defects Liability Period in the Main Contract and meet any reasonable expense of the Contractor caused by these defects. Provided that

   a reasonable examination by the Contractor would have foreseen the defects before fixing,
   and
   the defects are confined to defective materials and workmanship and not caused by the Contractor during storage or fixing, or by any other persons.

   36.4.3 Delivery to be in accordance with the agreed programme or, in the absence of a programme, at the reasonable request of the Contractor.

   36.4.4 A 5 per cent cash discount allowed to the Contractor for payment in full within 30 days of the end of the month in which delivery is made.

   36.4.5 No obligation to deliver after determination of the Contractor's employment (except where payment has been made in full).

36.4.6 Full payment to be made by the Contractor as described in 36.4.4 above.

36.4.7 Ownership passes to the Contractor upon delivery, whether or not paid for in full.

36.4.8 Disputes between Contractor and Nominated Supplier which are more or less relevant to disputes between the Employer and the Contractor which go to arbitration, will have the Arbitrator's award binding upon all parties.

36.4.9 Nothing in the Contract of Sale can supersede or override any of the foregoing sub-clauses.

36.5 Subject to Clauses 36.5.2 and 36.5.3 following, where the Contract of Sale in any way restricts, limits or excludes the liability of the Nominated Supplier to the Contractor, then the Contractor's liability to the Employer is restricted, limited or excluded to the same extent, but only if the Architect SPECIFICALLY APPROVES IN WRITING these restrictions, limitations or exclusions.

36.5.2 The Contractor can reject the nomination if this approval in writing from the Architect is not given. Any restrictions, limitations or exclusions will not affect the provisions of the Contract of Sale in Clause 36.4, but must be outwith them.

36.5.3 Clause 36.5 will not allow the Architect to nominate a supplier other than as set out in Clause 36.4.

Following is document Tender TNS/1 which is JCT Standard Form of Tender by Nominated Supplier incorporating Schedule 3 Warranty by a Nominated Supplier TNS/2. Immediately following the form are explanatory notes. The form itself makes quite clear whose responsibility it is to complete the relevant entries.

JCT

JCT Standard Form of Tender by Nominated Supplier

For use in connection with the Standard Form of Building Contract (SFBC) issued by the Joint Contracts Tribunal, 1980 edition, current revision

Job Title:
(name and brief location of Works)

[a] To be completed by or on behalf of the Architect/Supervising Officer.

Employer:[a]

Main Contractor:[a]
(if known)

Tender for:[a]
(abbreviated description)

ARCHITECT

Name of Tenderer:

To be returned to:[a]

[b] To be completed by the supplier; see also Schedule 1, item 7.

Lump sum price:[b] £

_____ (words)

and/or Schedule of rates (attached)

1 We confirm that we will be under a contract with the Main Contractor:

·1 to supply the materials or goods described or referred to in **Schedule 1** for the price and/or at the rate set out above; and

·2 in accordance with the other terms set out in that Schedule, as a Nominated Supplier in accordance with the terms of SFBC clause 36·3 to ·5 (as set out in **Schedule 2**) and our conditions of sale in so far as they do not conflict with the terms of SFBC clause 36·3 to ·5 [c]

[c] By SFBC clause 36·4·9 none of the provisions in the contract of sale can override, modify or affect in any way the provisions incorporated from SFBC clause 36·4 in that contract of sale. Nominated Suppliers should therefore take steps to ensure that their sale conditions do not incorporate any provisions which purport to override, modify or affect in any way the provisions incorporated from SFBC clause 36·4.

provided:

·3 the Architect/Supervising Officer has issued the relevant nomination instruction (a copy of which has been sent to us by the Architect/Supervising Officer); and

·4 agreement on delivery between us and the Main Contractor has been reached as recorded in **Schedule 1** Part 6 (see SFBC clause 36·4·3); and

·5 we have thereafter received an order from the Main Contractor accepting this tender.

2 We agree that this Tender shall be open for acceptance by an order from the Main Contractor within [d] of the date of this Tender. Provided that where the Main Contractor has not been named above we reserve the right to withdraw this Tender within 14 days of having been notified, by or on behalf of the Employer named above, of the name of the Main Contractor.

[d] May be completed by or on behalf of the Architect/Supervising Officer; if not so completed, to be completed by the supplier.

[e] To be struck out by or on behalf of the Architect/Supervising Officer if no Warranty Agreement is required.

3[e] Subject to our right to withdraw this Tender as set out in paragraph 2 we hereby declare that we accept the Warranty Agreement in the terms set out in **Schedule 3** hereto on condition that no provision in that Warranty Agreement shall take effect unless and until

a copy to us of the instruction nominating us,
the order of the Main Contractor accepting this Tender, and
a copy of the Warranty Agreement signed by the Employer

have been received by us.

For and on behalf of

Address

Signature   Date

© 1980 RIBA Publications Ltd

Tender TNS/1

Schedule 1

1. Description, quantity and quality of materials or goods:

   1A

   1B

   1C

Note: 1A to be completed by or on behalf of the Architect/Supervising Officer setting out his requirements. If the supplier is unable to comply with 1A he is to state in 1B what modifications he proposes, and the Architect/Supervising Officer is to state in 1C if such modifications are acceptable.

2. Access to Works:

   2A

   2B

*Note: 2A to be completed by or on behalf of the Architect/Supervising Officer. The supplier in 2B **either** confirms that the access in 2A is acceptable **or** states what modifications etc. to the access he requires **or** if 2A has not been completed, completes 2B.*

3. Provisions, if any, for returnable packings:

4. Completion Date of Main Contract (or anticipated Completion Date if Main Contract not let):

Note: To be completed by or on behalf of the Architect/Supervising Officer.

5. Defects Liability Period of the Main Contract   months.

Note: To be completed by or on behalf of the Architect/Supervising Officer.

Page 2

6A Anticipated commencement and completion dates for the nominated supply after any necessary approval of drawings (subject to SFBC clause 36·4·3, which provides that delivery shall be commenced, carried out and completed in accordance with any delivery programme agreed between the Contractor and supplier or in the absence of such programme in accordance with the reasonable directions of the Contractor):

6B ·1   Supplier's proposed delivery programme to comply with 6A:

·2   If 6B·1 not completed, delivery programme shall be to the reasonable directions of the Contractor.

6C Delivery programme as agreed between the supplier and the Contractor, if different from 6B:

Note: 6A to be completed by or on behalf of the Architect/Supervising Officer. The supplier to complete 6B·1 to take account of 6A; but the completion of 6B·1 is subject to the terms of 6C which may need to be used when the Contractor and supplier are settling item 6.

7.   Provisions, if any, for fluctuations in price or rates:

7A

7B

7C

Note: 7A to be completed by or on behalf of the Architect/Supervising Officer. If the supplier is unable to comply with 7A he is to state in 7B what modifications to the provisions in 7A he requires, and the Architect/Supervising Officer is to state in 7C if such modifications are acceptable.

8.   SFBC clause 25 (extensions of time) applies to the Main Contract without modification except as stated below:

The liquidated and ascertained damages (SFBC clause 2·4·2 and Appendix entry) under the Main Contract are at the rate of £ _____ per _____ .

Note: To be completed by or on behalf of the Architect/Supervising Officer.

9.   Contract of sale with Contractor to be under hand/under seal.

Note: Alternative not to be used to be deleted by the supplier subject to agreement on the method of execution of the sale contract with the Main Contractor.

Tender TNS/1

Schedule 2

JCT Standard Form of Building Contract

Clause 36·3 to ·8 provides as follows:

Ascertainment of costs to be set against prime cost sum

36·3 ·1 For the purposes of clause 30·6·2·8 the amounts "properly chargeable to the Employer in accordance with the nomination instruction of the Architect/Supervising Officer" shall include the total amount paid or payable in respect of the materials or goods less any discount other than the discount referred to in clause 36·4·4·, properly so chargeable to the Employer and shall include where applicable:

·1·1 any tax (other than any value added tax which is treated, or is capable of being treated, as input tax (as referred to in the Finance Act 1972) by the Contractor) or duty not otherwise recoverable under this Contract by whomsoever payable which is payable under or by virtue of any Act of Parliament on the import, purchase, sale, appropriation, processing, alteration, adapting for sale or use of the materials or goods to be supplied; and

·1·2 the net cost of appropriate packing carriage and delivery after allowing for any credit for return of any packing to the supplier; and

·1·3 the amount of any price adjustment properly paid or payable to, or allowed or allowable by the supplier less any discount other than a cash discount for payment in full within 30 days of the end of the month during which delivery is made.

·2 Where in the opinion of the Architect/Supervising Officer the Contractor properly incurs expense, which would not be reimbursed under clause 36·3·1 or otherwise under this Contract, in obtaining the materials or goods from the Nominated Supplier such expense shall be added to the Contract Sum.

Sale contract provisions – Architect's/Supervising Officer's right to nominate supplier

36·4 Save where the Architect/Supervising Officer and the Contractor shall otherwise agree, the Architect/Supervising Officer shall only nominate as a supplier a person who will enter into a contract of sale with the Contractor which provides, inter alia:

·1 that the materials or goods to be supplied shall be of the quality and standard specified provided that where and to the extent that approval of the quality of materials or of the standards of workmanship is a matter for the opinion of the Architect/Supervising Officer, such quality and standards shall be to the reasonable satisfaction of the Architect/Supervising Officer;

·2 that the Nominated Supplier shall make good by replacement or otherwise any defects in the materials or goods supplied which appear up to and including the last day of the Defects Liability Period under this Contract and shall bear any expenses reasonably incurred by the Contractor as a direct consequence of such defects provided that:

·2·1 where the materials or goods have been used or fixed such defects are not such that reasonable examination by the Contractor ought to have revealed them before using or fixing;

·2·2 such defects are due solely to defective workmanship or material in the materials or goods supplied and shall not have been caused by improper storage by the Contractor or by misuse or by any act or neglect of either the Contractor, the Architect/Supervising Officer or the Employer or by any person or persons for whom they may be responsible or by any other person for whom the Nominated Supplier is not responsible;

·3 that delivery of the materials or goods supplied shall be commenced, carried out and completed in accordance with any delivery programme agreed between the Contractor and the Nominated Supplier or in the absence of such programme in accordance with the reasonable directions of the Contractor;

·4 that the Nominated Supplier shall allow the Contractor a discount for cash of 5 per cent on all payments if the Contractor makes payment in full within 30 days of the end of the month during which delivery is made;

·5 that the Nominated Supplier shall not be obliged to make any delivery of materials or goods (except any which may have been paid for in full less only any discount for cash) after the determination (for any reason) of the Contractor's employment under this Contract;

·6 that full discharge by the Contractor in respect of payments for materials or goods supplied by the Nominated Supplier shall be effected within 30 days of the end of the month during which delivery is made less only a discount for cash of 5 per cent if so paid;

·7 that the ownership of materials or goods shall pass to the Contractor upon delivery by the Nominated Supplier to or to the order of the Contractor, whether or not payment has been made in full;

·8 that if any dispute or difference between the Contractor and Nominated Supplier raises issues which are substantially the same as or connected with issues raised in a related dispute between the Employer and the Contractor under this Contract then, where * articles 5·1·4 and 5·1·5 apply, such dispute or difference shall be referred to the Arbitrator appointed or to be appointed pursuant to article 5 who shall have power to make such directions and all necessary awards in the same way as if the procedure of the High Court as to joining one or more defendants or joining co-defendants or third parties was available to the parties and to him and in any case the award of such Arbitrator shall be final and binding on the parties:

·9 that no provision in the contract of sale shall override modify or affect in any way whatsoever the provisions in the contract of sale which are included therein to give effect to clauses 36·4·1 to 36·4·9 inclusive.

36·5 ·1 Subject to clauses 36·5·2 and 36·5·3, where the said contract of sale between the Contractor and the Nominated Supplier in any way restricts, limits or excludes the liability of the Nominated Supplier to the Contractor in respect of materials or goods supplied or to be supplied, and the Architect/Supervising Officer has specifically approved in writing the said restrictions, limitations or exclusions, the liability of the Contractor to the Employer in respect of the said materials or goods shall be restricted, limited or excluded to the same extent.

·2 The Contractor shall not be obliged to enter into a contract with the Nominated Supplier until the Architect/Supervising Officer has specifically approved in writing the said restrictions, limitations or exclusions.

·3 Nothing in clause 36·5 shall be construed as enabling the Architect/Supervising Officer to nominate a supplier otherwise than in accordance with the provisions stated in clause 36·4.

*The Architect/Supervising Officer should state whether in the Appendix to the SFBC the words "Articles 5·1·4 and 5·1·5" have been struck out; if so then clause 36·4·8 will not apply to the Nominated Supplier.

Page 4

Schedule 3 : Warranty by a Nominated Supplier

To the Employer:
named in our Tender dated

For:
(abbreviated description of goods/materials)

To be supplied to:
(job title)

1 Subject to the conditions stated in the above mentioned Tender (that no provision in this Warranty Agreement shall take effect unless and until the instruction nominating us, the order of the Main Contractor accepting the Tender and a copy of this Warranty Agreement signed by the Employer have been received by us) WE WARRANT in consideration of our being nominated in respect of the supply of the goods and/or materials to be supplied by us as a Nominated Supplier under the Standard Form of Building Contract referred to in the Tender and in accordance with the description, quantity and quality of the materials or goods and with the other terms and details set out in the Tender ('the supply') that:

1·1 We have exercised and will exercise all reasonable skill and care in:

1·1 ·1 the design of the supply insofar as the supply has been or will be designed by us; and

·2 the selection of materials and goods for the supply insofar as such supply has been or will be selected by us; and

·3 the satisfaction of any performance specification or requirement insofar as such performance specification or requirement is included or referred to in the Tender as part of the description of the supply.

1·2 We will:

1·2 ·1 save insofar as we are delayed by:

·1·1 force majeure; or

·1·2 civil commotion, local combination of workmen, strike or lock-out; or

·1·3 any instruction of the Architect/Supervising Officer under SFBC clause 13·2 (Variations) or clause 13·3 (provisional sums); or

·1·4 failure of the Architect/Supervising Officer to supply to us within due time any necessary information for which we have specifically applied in writing on a date which was neither unreasonably distant from nor unreasonably close to the date on which it was necessary for us to receive the same

so supply the Architect/Supervising Officer with such information as the Architect/Supervising Officer may reasonably require; and

·2 so supply the Contractor with such information as the Contractor may reasonably require in accordance with the arrangements in our contract of sale with the Contractor; and

·3 so commence and complete delivery of the supply in accordance with the arrangements in our contract of sale with the Contractor

that the Contractor shall not become entitled to an extension of time under SFBC clauses 25·4·6 or 25·4·7 of the Main Contract Conditions nor become entitled to be paid for direct loss and/or expense ascertained under SFBC clause 26·1 for the matters referred to in clause 26·2·1 of the Main Contract Conditions; and we will indemnify you to the extent but not further or otherwise that the Architect/Supervising Officer is obliged to give an extension of time so that the Employer is unable to recover damages under the Main Contractor for delays in completion, and/or pay an amount in respect of direct loss and/or expense as aforesaid because of any failure by us under clause 1·2·1 or 1·2·2 hereof.

Pages 1-4 comprising TNS/1 with Schedules 1 and 2 are issued in a separate pad.

© 1980 RIBA Publications Ltd

Warranty TNS/2

2   We have noted the amount of the liquidated and ascertained damages under the Main Contract, as stated in TNS Schedule 1, item 8.

3   Nothing in the Tender is intended to or shall exclude or limit our liability for breach of the warranties set out above.

4·1   If at any time any dispute or difference shall arise between the Employer or the Architect/Supervising Officer on his behalf and ourselves as to the construction of this Agreement or as to any matter or thing of whatsoever nature arising out of this Agreement or in connection therewith; then such dispute shall be and is hereby referred to the arbitration and final decision of a person to be agreed between the parties hereto, or, failing agreement within 14 days after either party has given to the other a written request to concur in the appointment of an arbitrator, a person to be appointed on the request of either party by the President or a Vice-President for the time being of the Royal Institute of British Architects.

4·2   ·1   Provided that if the dispute or difference to be referred to arbitration under this Agreement raises issues which are substantially the same as or connected with issues raised in a related dispute between the Employer and the Contractor under the Main Contract or between a Nominated Sub-Contractor and the Contractor under Agreement NSC/4 or NSC/4a or between the Employer and any Nominated Sub-Contractor under Agreement NSC/2 or NSC/2a, or between the Employer and any other Nominated Supplier, and if the related dispute has also been referred for determination to an arbitrator, the Employer and ourselves hereby agree that the dispute or difference under this Agreement shall be referred to the arbitrator appointed to determine the related dispute; and such arbitrator shall have power to make such directions and all necessary awards in the same way as if the procedure of the High Court as to joining one or more of the defendants or joining co-defendants or third parties was available to the parties and to him.

          ·2   Save that the Employer or ourselves may require the dispute or difference under this Agreement to be referred to a different arbitrator (to be appointed under this Agreement) if either of us reasonably considers that the arbitrator appointed to determine the related dispute is not properly qualified to determine the dispute or difference under this Agreement.

4·3   Paragraphs 4·2·1 and 4·2·2 hereof shall apply unless in the Appendix to the Main Contract Conditions the words 'Articles 5·1·4 and 5·1·5 apply' have been deleted.

4·4   The award of such arbitrator shall be final and binding on the parties.

Signature of Supplier: ....

Signature of Employer: ....

Explanatory notes

As already stated, the instructions for the use of the form are clearly set out against each entry to be made. Note that in certain respects there is a similarity to Tender NSC/1 in as much as some clauses are filled in by the Supplier as his intentions, and some by the Architect as his intentions. In all cases there is a further paragraph to allow changes or agreement to the intentions.

Page 1   The job title, Employer's name, and name of the Main Contractor, will be completed by the Architect. The Tender description, also completed by the Architect, in an abbreviated version, must contain information on any design or selection requirements, but this is enlarged upon in Schedule 1. The Tenderer enters his own name and the Architect the returning address.

The Tender amount can be entered as a lump sum price or as a schedule of rates which will be attached to the documents. In the second case, this would consist of a list of items without quantities which is rated by the Tenderer. These rates will be used to extend measured quantities of goods and materials supplied. Note here also the reference to Item 7 Schedule 1 where there is an agreement to be made in regard to the inclusion of fluctuations.

Item 1   The important points here are. that the Supplier confirms that the Contract is with the Main Contractor and that it refers only to the materials and goods in Schedule 1 and for the price offered, in whichever form. There is also confirmation that the offer abides with the other terms in Schedule 1, is in accordance with Clauses 36.3 to 36.5 of Schedule 2 and, most important, that the conditions of sale are not in conflict with the stated clauses. The sidenote to the form clearly establishes the position with regard to conditions of sale provisions.

The above confirmations are given only if certain provisions have been made, as follows:

(a) Nomination has been made by the Architect.
(b) A copy of the Nomination has been sent to the Supplier.
(c) Agreement with the Main Contractor on delivery has been made as recorded in Schedule 1.
(d) Acceptance of the offer has been made by the Main Contractor. (Note that acceptance by the Architect is not valid.)

Item 2   The Supplier can withdraw his Tender on two counts:

(a) Within 14 days after notification of the name of the Contractor. If his name is known at time of tendering, of course this will not apply.
(b) The Main Contractor's failure to accept the Tender within the period entered here. The time open for acceptance can be entered either by the Architect, or by the Supplier, if left blank.

Item 3   This item may be struck out by the Architect if no warranty is required. If a warranty is required, it is given here only if

(a) A copy of the nomination instruction has been received,
(b) The Main Contractor's acceptance has been received,

(c) A copy of the signed Warranty Agreement by the Employer has been received.

The Warranty gives the Employer direct legal rights against the Supplier for his liability as regards

(i) Design of the items supplied,
(ii) Any delay by the Nominated Supplier,
(iii) Any delay in providing information.

The name and address of the Supplier is inserted at the foot of the page and signed and dated, on his behalf, by a person entitled to do so.

Schedule 1

Item 1A   The abbreviated description on Page 1 is enlarged upon in detail. Lack of space may mean a separate sheet being attached, titled and signed by the Architect.

Item 1B   The Supplier has the opportunity to modify the requirements in 1A above, if he cannot meet their requirements.

Item 1C   The Architect states here his acceptance of these modifications, if he agrees with them. If agreement cannot be reached here between the Architect and the Supplier, nomination will not take place and, presumably another supplier will have to be approached.

Item 2   Access to the Works can be copied from either the Preliminaries Bill or the Appendix to the Main Contract. In 2B the Supplier either agrees to the access or suggests his own modifications, which might well be a greater width for wide loads, or any other requirement. The Architect is not obliged to complete 2A, but the Supplier has to complete 2B, in either case.

Item 3   The Supplier will state here if he wishes packaging to be returned, when and in what form, and the cash credit for the cost of return.

Item 4   This is copied from the Main Contract, either from the Appendix or the Preliminaries Bill.

Item 5   The above comments also apply here.

Item 6   This item should result in an agreed delivery programme between the Contractor and the Supplier. In 6A the Architect sets out the commencement and completion dates of delivery to suit the requirements of the Contract, which time scale has to include allowance for the approval of drawings, if any, and is subject to Clause 36.4.3. In 6B.1 the Supplier intimates his start and finish dates to comply with the Architect's requirements in 6A. If the Supplier does not enter any date in 6B.1, then he is tacitly agreeing to comply with a reasonable delivery programme requirement of the Contractor. 6C sets out the final agreement made, only if different from 6B.1 or 6B.2 above.

Item 7   This item refers to fluctuations and again is in three parts. 7A is the Architect's requirement, 7B the Supplier's modifications and 7C the

Architect's acceptance of these modifications. Of course, if 7A is acceptable to the Supplier, he will strike out 7B and 7C.

If fluctuations are to apply, it would be insufficient to quote Clauses 38, 39 or 40 as a description. Further information would be required, such as, tender date, basic list of materials, fuels option, base date, etc.

Item 8   Any modifications to Clause 25 in the Main Contract are stated here.

The information to be copied from the Main Contract Appendix with regard to Liquidated and Ascertained Damages.

Item 9   After agreement with the Contractor, the Supplier deletes whichever form he does not require in order to execute the Contract.

Schedule 2

This schedule is an extract from the Standard Form, Clauses 36.3 to 36.5, all of which have been commented on previously.

The footnote to this page points out that Clause 36.4.8 will not apply if Articles 5.1.4 and 5.1.5 have been struck out of the Main Contract. This is again reiterated in the Warranty under Clause 4.3.

The Architect might well strike out Clause 36.4.8 in Schedule 2 and sidenote to the effect that Articles 5.1.4 and 5.1.5 do not apply.

Schedule 3 – Warranty

The Warranty is signed by the Supplier and the Employer and is therefore an agreement between them. However, it is one-sided since the Employer does not agree to do anything, but the Nominated Supplier does.

Item 1   Before the Warranty becomes effective, the following must be satisfied:

(a) A tender has been submitted by the Supplier.
(b) The tender has not been withdrawn.
(c) A copy of the nomination instruction has been received.
(d) The Contractor has accepted the tender.
(e) The Architect has agreed to any proposed modifications of the Supplier, where he is obliged to do so.
(f) A programme has been agreed between the Contractor and the Supplier.

Item 1.1   The Nominated Supplier warrants for any design work, selection of goods and materials and satisfaction of any performance specification. This gives the Employer a legal right to recover through the Courts any breach by the Supplier in these respects. If no design element or selection of materials, etc. is required, the clause automatically ceases to be effective and no deletion is needed.

It now can be seen that it is important for the abbreviated description of the works, in Page 1 of the Tender to set out clearly all the above requirements.

Item 1.2   The Warranty also covers the Supplier's obligation to provide information to both the Architect and the Contractor, as the Architect may reasonably require and to the Contractor as may reasonably require, but in accordance with their arrangements in the Tender. The Supplier then warrants to commence and complete delivery in accordance with the agreed programme. These warranty matters, are however, subject to any delay caused by the events listed in Clauses 1.2.1.1 to 1.2.1.4 inclusive.

The Nominated Supplier further warrants that he will perform his obligations in order that the Contractor will not be entitled to an extension of time or a claim for direct loss and/or expense from the Employer, due to his non-performance. If the Supplier is late in passing information to the Architect, he in turn might be late in passing necessary information to the Contractor. This could give rise to a grant of extension of time and a possible reimbursement from the Employer for direct loss and/or expense.

If the Supplier defaults under 1.2 and an extension of time is granted and reimbursement made to the Contractor by the Employer, this Warranty gives the Employer the right to recover his loss from the Supplier.

The likely claim against the Supplier from the Employer would be made up as follows:

(a) Loss of Liquidated and Ascertained Damages for the period of the extension of time.
(b) Amount of reimbursement paid to the Contractor.

The dates agreed in the programme now become very important; they are the only basis on which a delay can be calculated, and the only check on the Supplier's non-performance in these matters.

Item 2   The Supplier agrees to the amount of the Liquidated and Ascertained Damages in the Main Contract by noting their insertion in Item 8 of Schedule 1.

Item 3   The Supplier agrees that nothing in his Tender will have any overriding effect on his Warranty.

Item 4   This is the standard clause with reference to arbitration, which has been fully explained elsewhere. Note, however, that the 'enjoinder clauses' will not apply if articles 5.1.4 and 5.1.5 have been deleted in the Main Contract.

The Warranty is signed by both the Supplier and the Employer. The Architect should not sign on the Employer's behalf as he is a party to this agreement, and therefore must sign for himself.

Scotland

Clause 36 – Nominated Suppliers

In Scotland the Scottish Building Contract which contains the Scottish Supplement makes one change and one addition to the above clause, namely:

Clause 36.4.8 is deleted and the clause changed to the Scottish term 'Arbiter' to be appointed in terms of Clause 4 of the Scottish Building Contract. Otherwise the wording is substantially the same.

Clause 36.6 is additional, that in the case of determination of the Contractor's employment the Nominated Supplier shall recognize an assignation by the Contractor in favour of the Employer in terms of Clause 27.4.2.1.

Also, in Scotland there is a green form titled 'Standard Form of Tender for use in Scotland by Nominated Suppliers'. This form is not mandatory for the Architect's use, but is advisable to ensure that the Tenderer's offer is in accordance with Clause 36.

A second green form titled 'Standard Form of Nomination of Suppliers for Use in Scotland' is also available but again not mandatory for the Architect's use.

Copies of the two forms follow, with the various parts indicated to show which party enters the information, together with explanatory notes on each form.

Instructions for Use

1. Before issuing a principal and copy to the Tenderer the Architect/Supervising Officer must

   (a) complete Page 1 (including 1.2 if desired)

   (b) delete 4 on Page 2 if a Warranty is not required

   (c) complete items 1–5 of the Schedule, and items 6–8 if desired

   (d) if a Warranty is required insert the names and addresses of the Supplier and the Employer on Page 5: if not required draw a line through each page of the Warranty.

2. The Tenderer completes Page 1 and the Schedule as required:

   Signs the Warranty on Page 6 (if required):

   Signs the tender on Page 2 and returns the principal to the Architect:

   Retains copy for his own use.

3. The Architect arranges for the Employer to sign the Warranty (if required).

4. The Architect sends principal of completed document to Main Contractor and photocopy of Warranty (if required) to Tenderer.

SBCC

STANDARD FORM OF TENDER

for use in Scotland

by

NOMINATED SUPPLIERS

_____ Supplier

_____ Job Title

_____ Employer

_____ Main Contractor

_____ Architect

The constituent bodies of the
Scottish Building Contract Committee are:

Royal Incorporation of Architects in Scotland

Scottish Building Employers Federation

Scottish Branch of the Royal Institution of
  Chartered Surveyors

Convention of Scottish Local Authorities

Federation of Specialists and Sub-Contractors
  (Scottish Board)

Committee of Associations of Specialist
  Engineering Contractors (Scottish Branch)

Association of Consulting Engineers
  (Scottish Group)

Copyright of the S.B.C.C., 39 Castle Street, Edinburgh

February 1980

Standard Form of Tender for use in Scotland by Nominated Suppliers

(For use when the Supplier is nominated under the Building Contract and Scottish Supplement to the Conditions of the Standard Form of Building Contract [Clause 36 of these Conditions refers])

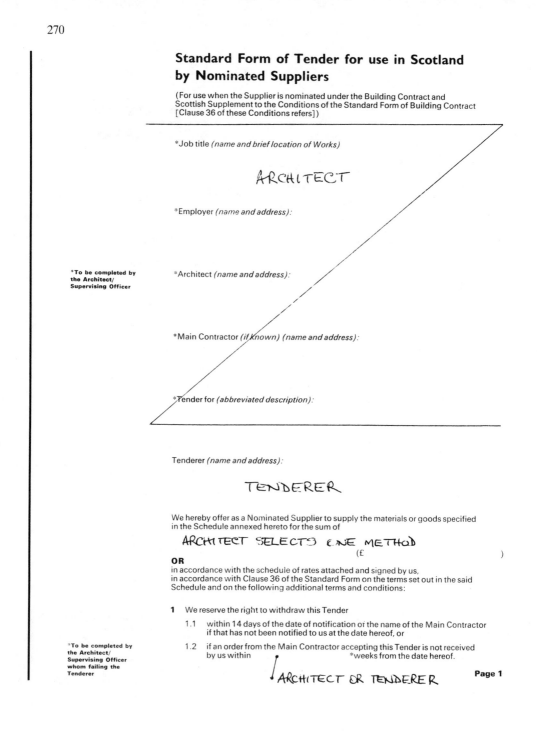

*Job title *(name and brief location of Works)*

ARCHITECT

*Employer *(name and address)*:

*To be completed by the Architect/ Supervising Officer

*Architect *(name and address)*:

*Main Contractor *(if known) (name and address)*:

*Tender for *(abbreviated description)*:

Tenderer *(name and address)*:

TENDERER

We hereby offer as a Nominated Supplier to supply the materials or goods specified in the Schedule annexed hereto for the sum of

ARCHITECT SELECTS ONE METHOD

(£   )

OR

in accordance with the schedule of rates attached and signed by us,
in accordance with Clause 36 of the Standard Form on the terms set out in the said Schedule and on the following additional terms and conditions:

1   We reserve the right to withdraw this Tender

   1.1   within 14 days of the date of notification of the name of the Main Contractor if that has not been notified to us at the date hereof, or

   1.2   if an order from the Main Contractor accepting this Tender is not received by us within   *weeks from the date hereof.

*To be completed by the Architect/ Supervising Officer whom failing the Tenderer

ARCHITECT OR TENDERER

Page 1

2 This Tender shall not become binding on us until

2.1 the Architect/Supervising Officer has issued a nomination instruction to the Main Contractor and we have received a copy thereof from the Architect/Supervising Officer, and

2.2 we have reached agreement with the Main Contractor on delivery in accordance with Clause 36.4.3 of the Standard Form

notwithstanding any acceptance by the Employer or Architect/Supervising Officer on his behalf.

3 We bind ourselves to recognise an Assignation by the Main Contractor in terms of Clause 27.4.2 of the Standard Form of Building Contract.

4* We undertake to enter into the Warranty in favour of the Employer annexed hereto if so requested by the Architect/Supervising Officer, subject to prior compliance with Conditions 1 and 2 hereof.

*To be deleted by the Architect/Supervising Officer if not required

Signed by or on behalf of the Nominated Supplier_____

Date_____

Schedule

The following are to be completed by the Architect/Supervising Officer prior to issue:

1. Description, Quantity and
   Quality of Materials and
   Goods

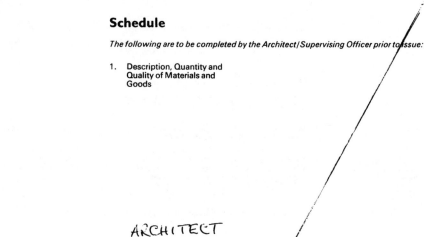

ARCHITECT

2. Completion Date, or anticipated
   Completion Date, of Main Contract

3. Defects Liability Period of
   Main Contract   _____months

4. Clause 25 (Extensions of Time)
   of the Standard Form applies
   to the Main Contract in full
   unless noted otherwise

***To be completed only
when a warranty
is required**

5.* Liquidate and ascertained
   damages in the Main Contract   £_____ per_____

The following may be completed by the Architect/Supervising Officer but if not so completed then by the Tenderer:

6.   Access to Works

ARCHITECT OR TENDERER

7.   Anticipated commencement date
     and completion date of the
     nominated supply, and any terms
     relating to delivery
     *(see Clause 36.4.3 of the
     Standard Form)*

8.   Provisions, if any, for
     fluctuations in prices or
     rates

The following is to be completed by the Tenderer:

9.   Provisions, if any, for
     returnable packagings

Warranty

by

(hereinafter referred to as 'the Supplier')

in favour of

(hereinafter referred to as 'the Employer')

ARCHITECT

CONSIDERING THAT the Supplier has submitted a Tender dated
for the supply of
and that the said Tender has been approved by the Architect/Supervising Officer
therein referred to, on behalf of the Employer
FURTHER CONSIDERING that the Supplier as one of the conditions of the said
Tender has agreed to enter into this Warranty T H E R E F O R E, subject always to
the terms of Conditions 1 and 2 of the said Tender, the Supplier HEREBY WARRANTS
to the Employer

1.   We have exercised and will exercise all reasonable skill and care in

   1.1   the design of the materials and goods supplied so far as the same have been
         designed by us in accordance with any performance specification or
         requirement included or referred to in the Tender and

   1.2   the selection of materials and goods supplied in so far as the same have been
         or will be selected by us.

2.   We will supply the Architect/Supervising Officer or the Contractor with such
     information as either may reasonably require and commence and complete
     delivery in accordance with the agreed delivery programme so that the
     Contractor will not be entitled

   2.1   to an extension of time under Clauses 25.4.6 or 25.4.7 of the Standard Form,
         or

   2.2   to be paid for direct loss and/or expense under Clause 26.1 for the matters
         referred to in Clause 26.2.1 of the Standard Form.

3.   In the event of the Main Contractor becoming entitled under Clauses 2.1 or 2.2
     hereof we undertake to indemnify the Employer to the extent that the Employer
     is obliged to grant the Main Contractor an extension of time and be unable to
     recover damages from the Main Contractor for delay in completion and/or pay the
     Main Contractor an amount in respect of direct loss and/or expense.

4. Any dispute or difference of opinion arising out of the terms and conditions of this Warranty shall be referred to the arbitration of a person to be agreed between us and the Employer, or failing agreement within 14 days of either party sending written notice to the other requesting arbitration by an Arbiter appointed by the Sheriff of the Sheriffdom in which any of the materials or goods to be supplied are to be delivered.

4.1 The Arbiter shall without prejudice to the generality of his powers have power to make such orders and directions as may in his opinion be necessary to determine the rights of the parties, and to open up review and revise any directions requirements or notices given by either party to the other, and to determine all matters in dispute as if no such decision requirement or notice had been given.

4.2 The Law of Scotland shall apply to all arbitrations under these presents and the award of the Arbiter shall be final and binding on the parties subject always to the provisions of Clause 3 of the Administration of Justice (Scotland) Act 1972.

4.3 The Arbiter shall be entitled to remuneration and reimbursement of his outlays.

5. This Warranty shall be construed and the rights of the Supplier and the Employer and all matters arising hereunder shall be determined according to the Law of Scotland: IN WITNESS WHEREOF these presents are executed at
on the
day of   19   before these witnesses subscribing.

_____witness

_____address

_____occupation

                              } _____Supplier.

_____witness

_____address

_____occupation

_____witness

_____address

_____occupation

                              } _____Employer.

_____witness

_____address

_____occupation

N.B.—This document is set out as for execution by individuals or firms: Where Limited Companies or Local Authorities are involved amendments will be necessary and the appropriate officials should be consulted.

Explanatory notes

This form, although not mandatory, does incorporate all the provisions of Clause 36 with special emphasis on the necessary requirements of Clause 36.4 which are very important qualifications as regards nomination. Remember that the Architect can deviate from Clause 36.4 requirements only if he has a specific agreement with the Contractor to do so.

The instructions for the use of the form are clearly set out on the last page and give the order of procedure for the Architect and the Tenderer. In his own and his clients interests the Architect would be well advised to obtain tenders using this form, as it acts as a check list to enable a correct and viable nomination to be made.

Page 1   The first five headings are completed by the Architect whilst the Tenderer will enter his own name and address. An abreviated description of the Tender subject is sufficient as a detailed list is completed in the Schedule. Notwithstanding, the Tender subject will be clear and concise and contain information on any design or selection requirements.

Near the bottom of the page is an alternative for the type of tender basis required. Either could be used, but one should be chosen and the other struck out. The first alternative, the 'Schedule', would refer to a list of items with quantities which can be rated, extended and totalled to give a lump sum tender. The second alternative, the 'Schedule of Rates', would be a list of items without quantities which is rated by the Tenderer. These rates will be used to extend measured quantities of goods and materials supplied.

Bear in mind that the Main Contractor has items in his Bill for the fix-only of the materials and goods, which includes for taking delivery, unloading, storage, handling, hoisting and returning packages if required. Therefore, the Nominated Supplier will include in his Tender for delivery only to the site.

Clause 1   The Supplier can withdraw his Tender on two counts:

    (a) Within 14 days after notification of the name of the Contractor. If his name is known at time of tendering, of course this will not apply.

    (b) The Main Contractor failing to accept the Tender within the period entered here.

Clause 2   The important point to arise here is that acceptance of the Tender is made *only* by the Main Contractor. The clause states that the Employer's or Architect's acceptance has no validity.

The Clause sets out the conditions to be satisfied before the Tender becomes binding on the Supplier, namely:

    (a) Architect has issued a nomination with a copy to the Supplier.

    (b) Contractor and Supplier have reached agreement on delivery programmes.

Clause 3   With all Scottish sub-contract forms this assignation clause is inserted to bring the Sub-contract into line with the Main Conditions where, under Clause 27.4.2, the Contractor agrees to assign the benefits of any arrangements for the supply of goods, etc., for the Contract, to the Employer, in the case of the determination of the Contractor's employment (but not by bankruptcy, etc.). This clause then binds the Supplier to do likewise.

Clause 4   This clause can be either left in or deleted, dependent upon the need for a warranty. If a warranty is required and the clause is therefore left intact, it gives the Employer direct legal rights against the Supplier for his liability, as regards:

(a) Design of the items supplied,
(b) Any delay by the Nominated Supplier, and
(c) Any delay in providing information.

The Warranty is attached to the Tender form and sets out the points of guarantee that the Tenderer will give.

Schedule

Item 1   Here the abbreviated description on Page 1 is enlarged upon in detail. If lack of space on the form is a problem, a separate sheet can be attached to the form, suitably titled and signed to form part of the Tender.

Item 2   This is copied from the Main Contract, either from the Appendix II or the Preliminaries Bill.

Item 3   The above comments apply here also.

Item 4   This would be standard unless any amendments had been made in the Main Contract, which would have to be noted here.

Item 5   This information would only be given if a Warranty is required and it is copied from the Main Contract. The Contractor can receive an extension of time under the Main Contract due to the Supplier being late in providing information and that extension precludes the Employer from applying Liquidated and Ascertained Damages for that period of time. This is then a loss to the Employer.

Item 6   All the information in the Preliminaries Bill relating to access and any restrictions imposed, would be entered here.

Item 7   These dates will allow an agreement to be made as to programme between the Contractor and the Nominated Supplier. (*See also* comments on Clause 36.4.3.)

Item 8   Although some form of fluctuations apply to the Main Contract, it may be, due to the short duration for the supply of the materials, that fluctuations are not necessary. In this case the Architect will state that no fluctuations will apply. On the Supplier's part, he will have to include in his Tender for any foreseeable increases.

If fluctuations are to apply, it would be insufficient to quote Clauses 38, 39 or 40 as a description. The further applicable information is required, such as, tender date, basic list of materials, fuels option, base date, etc.

Item 9   The Tenderer will state here if he wishes packaging to be returned, when and in what form, and the cash credit for the cost of return.

Warranty

This form of Warranty will be used only if Clause 5 is not deleted. The Warranty is signed by the Nominated Supplier and the Employer and therefore it is an agreement between them. However, it is one-sided since the Employer does not agree to do anything but the Nominated Supplier does.

Considerations have to be satisfied before the Warranty will become effective, namely:

(a) That the Supplier has submitted a tender,
(b) That on the Employer's behalf the Tender has been approved by the Architect. (An approval can be written in a convenient space, such as Page 2, signed and dated by the Architect.),
(c) Clause 1 – Tender has not been withdrawn and has been accepted by the contractor.
    Clause 2 – Architect has issued nomination with a copy to the Supplier.
    Contractor and Supplier have agreed programme.

Clause 1   The Nominated Supplier here warrants for any design work to accord with any performance specification and selection of goods and materials. This gives the Employer a legal right to recover through the Courts any breach by the Supplier in these respects.

If no design element or selection of materials is required, the Clause automatically ceases to be effective and no deletion is necessary.

With the Warranty in mind, the abbreviated description of the Tender on Page 1 must clearly set out any design or material selection requirements.

Clause 2   Here the Nominated Supplier warrants that he will supply the Architect or the Contractor with required information and commence and complete the agreed delivery programme to the agreed dates. He will further warrant to perform these obligations, in order that the Contractor will not be entitled to an extension of time or a claim for direct loss and/or expense from the Employer, due directly to his non-performance. If the Supplier is late in passing required information to the Architect, the latter also might be late in passing necessary information to the Contractor. This could give rise to an extension of time being granted to the Contractor under Clause 25.4.6 and could also entitle him to a claim for direct loss and/or expense from the Employer.

Clause 3   When the Supplier fails under Clause 2 and an extension of time and reimbursement is made to the Contractor by the Employer, this part of the Warranty gives the Employer recoverable rights against the Supplier. The Employer's loss is the right to deduct Liquidated and Ascertained Damages from the Contractor, which is stated as an amount per week in Item 5 of Page 3 of the Tender. It is calculated by multiplying the amount per week by the length of time of the extension granted.

Thus the Employer's loss added to that of the Contractor's grant of loss and expense paid by the Employer, will be the total likely claim levied at the Supplier.

It is also possible under Clause 25.4.7 for delay on the part of the Supplier to cause an extension of time to be granted to the Contractor. Again the Employer loses his right to deduct Liquidate and Ascertained Damages, thus sustaining a loss recoverable under this Warranty. Item 7 of the Schedule, showing start and finish dates and agreed programme, would form the basis for the decision as to whether or not a delay had been incurred.

Clauses 4 and 5 These are the standard clauses with regard to disputes being referred to arbitration and that the law of Scotland will operate, and they have been fully explained elsewhere.

Again, with regard to the signing of the document, it would be advisable to seek legal advice. Note the footnote at the bottom of Page 6.

280

Standard Form for **Nomination** of **Suppliers** for Use in **Scotland**

(For use when the Supplier is nominated under the Building Contract and Scottish Supplement to the Conditions of the Standard Form of Building Contract. Clause 36 of these Conditions applies).

To: Main Contractor *(name and address)*

Job Title *(name and brief location of Works)*

Prime cost or provisional sums on the following page(s) of Bills or Specification

Nominated Supplier *(name and address)*

ARCHITECT

Tender: Amount: £   Dated:
OR
Schedule of Rates (annexed to Tender)   Dated:

1   The Nominated Supplier has submitted a tender on the Standard Form of Tender for use in Scotland by Nominated Suppliers to which the above mentioned prime cost or provisional sums relate.

2   In accordance with Clause 36.2 of the Main Contract the above named is hereby nominated as a Supplier under the Main Contract.

3   The said Tender/Schedule together with any other relevant documents is enclosed.

*Delete as required

4   The Nominated Supplier has*/has not undertaken to enter into the Warranty annexed to the Standard Form of Tender for use in Scotland by Nominated Suppliers.

5   The Tender is to be accepted by you within the period stipulated in Clause 1.2 of the Tender.

6   In accordance with Clause 36.5 of the Main Contract I/we hereby approve the restrictions limitations and exclusions (if any) included in the Tender submitted by the Nominated Supplier.

_____Architect/Supervising Officer

_____Address

_____Date

List of relevant documents attached:_____

_____

_____

Issued to: Main Contractor	Q.S.	Clerk of Works (unpriced copy)	Nominated Supplier	Consulting Engineer	A/S.O. file
☐	☐	☐	☐	☐	☐

Copyright SBCC, 39 Castle Street, Edinburgh   February 1980

Explanatory notes

This (green-coloured) form is not mandatory for the Architect's use in making a nomination. However, it is surely more convenient for the Architect to use a standard form rather than write his own nomination instruction.

The form reiterates a great deal of the information contained in the tender and in particular states

(a) The basis of the Tender amount; one alternative to be deleted.
(b) Whether or not the Warranty has been entered into.
(c) Whether or not any restrictions, limitations and exclusions have been approved by the Architect (*see* comments on Clause 36.5 of the Main Conditions).

For the Supplier's Tender to become binding, a copy of this Nomination Instruction must be issued to the Supplier as part of the procedure.

Chapter 14

Completion

Clause 17 – Practical Completion and Defects Liability
Clause 18 – Partial possession by Employer
England and Wales

Clause 17 – Practical Completion and Defects Liability

There is no real definition of 'Practical Completion' set out in the Conditions and the only person empowered to decide that Practical Completion has taken place, or has been arrived at, is the Architect. The Architect issues a certificate naming the date on which, in his opinion, Practical Completion was achieved. This date could of course be reviewed by an Arbitrator. The Contractor himself might, however, decide that Practical Completion is close and write to the Architect suggesting that Practical Completion will be achieved at a stated time. There is no contractural obligation on him to do so, but it could have the effect of reminding the Architect and so start proceedings towards the Final Account that much more quickly. If the Employer takes possession of the works, then by inference Practical Completion must have taken place, but possession must not be allowed without the Certificate of Practical completion being issued to the Contractor.

17.1	The certificate and its date are very important, as they cause a series of clauses either to cease or to commence.

Cessation

Article 3	End of waiting period for arbitration over disputes.
Clause 13.2	End of the obligation to carry out instructions requiring a variation, although there could still be instructions regarding variations in progress or the sanction of a variation already raised.
Clause 17.5	End of liability for frost damage except that damage caused by frost prior to Practical Completion.
Clause 22A.1	End of obligation for insurance against Clause 22 Perils, if applicable.
Clause 24.2	End of liability for Liquidated and Ascertained Damages.

Clause 30.1.3   End of normal monthly interim certificates.

Clause 30.4.1   End of full retention on certificates; retention is halved.

Commencement

Article 3   Start of immediate arbitration over disputes.

Clause 17.2   Start of Defects Liability Period.

Clause 30.1.3   Start of halved retention percentage.

Clause 30.6.1.1 Start of Period of Final Measurement and Valuation, if stated in the Appendix.

Start of the period for the production of documents by the Contractor for adjustment of the Contract Sum (or before Practical Completion).

17.2   The issue of the Certificate of Practical completion has now caused the Defects Liability Period to commence and puts an obligation upon the Contractor to make good defects. These defects are

(1) Workmanship or materials not up to the standard of the Contract.
(2) Damage caused by frost which occurred before the date of Practical Completion.

The Architect will issue a Schedule of Defects, sometimes known as a 'snagging list', which is an instruction. He has the whole of the Defects Liability Period and 14 days beyond its finish date, in which to issue this instruction. In practical terms an Architect would probably wish to issue his Schedule of Defects as early as possible, in case of any major defects or a very large number of minor defects coming to light after his inspection.

The Contractor is contractually bound to obey the Instruction and carry out the work in the Schedule.

It is, of course, not improper for the Contractor to disagree with any individual items listed in the Schedule of Defects. If he thinks that any of the defects are not due to sub-standard materials or workmanship, or are frost damage occurring after Practical Completion, he can challenge the Architect immediately in writing. He would state the items in dispute and his reasons, and request a meeting. Obviously, if the Architect retracts and agrees to omit these items, then the Contractor does not have to make them good. If the suggestion is made that he does carry out the work then it would have to be on some other agreed cost basis, rather than on the normal valuation procedure.

Remember that the Architect has issued an instruction, and that failure on the Contractor's part to carry it out after a seven days notice requiring him to do so, will result in the Architect taking action. The possible course of this action has been explained previously.

17.3   This clause allows the Architect to issue instructions with regard to the making good of defects at any time before the Defects Liability Period

ends, and there is no restriction as to the number of them. These
instructions are limited only by the issue of the Schedule of Defects
because no other instructions in regard to defects are valid after the
schedule has been sent to the Contractor. However, as the Schedule of
Defects can be as late as 14 days after the end of the Defects Liability
Period, instructions could be issued for the whole length of that
period.

Again, you are bound to carry out the work and make good the
defects.

17.4   The Architect forms his own opinion as to when all defects are made
good to his reasonable satisfaction and at that time issues a Certificate
of Completion of Making Good Defects. This certificate has the
following effects:

(a) **30.4** – Retention percentage is now reduced to NIL.
(b) **30.6** – End of period during which the Contractor produces
documents for the adjustment of the Contract Sum.
(c) **30.8** – A pre-condition to the issue of the Final Certificate.

An important point for the Contractor to note is that, notwithstanding the issue of this Certificate, he is still responsible for defects up to
the issue of the Final Certificate and beyond – *see* Limitation Act 1980
and Prescription and Limitation (Scotland) Act 1973. The Architect
does not have the power to issue instructions after his Schedule of
Defects or 14 days after the end of the Defects Liability Period,
whichever comes first, but he may still bring any other defects to the
attention of the Contractor, who would be well advised to make them
good.

17.5   Here it is reinforced that the Contractor has no responsibility to make
good defects caused by frost damage when the frost occurred after the
date of Practical Completion, but conversely he is responsible when
the frost occurred before the date.

Diagram 11 opposite is an illustration of the effect on Practical
Completion of various happenings.

Clause 18 – Partial Possession by Employer

This clause envisages the Employer wishing to take over part of a building which
has been completed by the Contractor. It has nothing whatsoever to do with
sectional completion which is pre-determined at tendering stage and set out in both
the Bills and the Contract Conditions. In this instance it could be rooms, a suite of
rooms, a floor, or an entire block, any of which is complete and which the
Employer wishes to occupy for any number of reasons.

18.1   The clause cannot operate nor can the Employer have possession if the
Contractor does not give his consent. It may be that the giving of
partial possession would upset his site management arrangements or
his programme of work, and these objections would be intimated to
the Architect. However, it could be that objections can be overcome

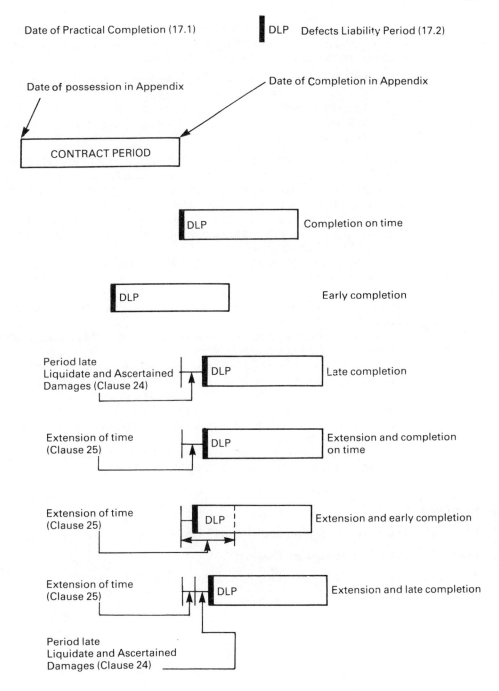

Diagram 11: Practical Completion

or that the Contractor's consent could be unconditional in the first place. Consent will be in writing as it is a notice to the Architect and could be to the Contractor's advantage, thus

(a) Practical Completion is deemed to have taken place for that part, and the effects in Clause 17 thus refer.
(b) Insurance under Clause 22A can be reduced.
(c) Liquidated and Ascertained Damages are reduced by a proportional amount.

The Architect is obliged to issue a certificate, sometimes called the 'Certificate of Partial Practical Completion', which states the estimate of the value of the relevant part. Note that no date of possession of the relevant part is stated, but it is advisable to record it.

Although there is a certificate issued, no provision is made for a separate period of Final Measurement and Valuation, nor for a separate Final Certificate.

It is only insurance under Clause 22A that would be reduced, but be careful, the value of the relevant part is obviously less than the full reinstatement value together with professional fees, if applicable. Consultation with the insurance company would be advisable before making a reduction of cover.

The Employer, on partial possession, takes complete responsibility for insurance against Clause 22 Perils.

The amount to be allowed for Liquidated and Ascertained Damages is reduced proportionately by taking the value of the relevant part in the Bill to the Contract Sum and reducing the damages by the same proportion. A careful check must be made here; the value of the relevant part in the certificate may well be different from the value of that part in the Bill of Quantities.

As stated in Clause 17, possession should not be permitted without a relevant certificate, and this is applicable to Clause 18 also.

Scotland

Clause 17 – Practical Completion and Defects Liability

Clause 18 – Partial Possession by Employer

No changes are required in the above clauses for use in Scotland.

Chapter 15

Non-completion

Clause 24 – Damages for Non-completion

Clause 27 – Determination by the Employer

Clause 28 – Determination by the Contractor

England and Wales

Clause 24 – Damages for non-completion

24.1 If the Contractor has gone beyond the completion date stated in the Appendix, taking into account all grants of extension of time, and Practical Completion has not been achieved or certified by the Architect, then the Architect will issue a certificate to the Employer (copy to the Contractor) to that effect. In practical terms the certificate means that the Contractor is late in his completion of the works. Until this certificate is issued, the Employer has no rights to Liquidated and Ascertained Damages.

24.2 The issue of the above certificate begins the Contractor's liability to the Employer for Liquidated and Ascertained Damages. The amount that can be claimed is set out in the Appendix, as a sum per week, stated either as NIL or as 'a sum calculated at the rate stated in the Appendix as liquidated and ascertained'. This latter wording would mean that the Contractor is responsible for the actual damages sustained by the Employer, which he would have to prove.

The amount stated per week as damages has to be reasonable and have a direct relationship to the Employer's probable loss. If the Contractor is uncertain or thinks that the damages are unrealistic, the time to challenge them is at tendering stage. By signing the Contract the Contractor has then accepted the amount as being a realistic pre-estimate and probably could not have the amount changed even in the Courts. It is possible that the Employer might not incur the full amount of damages as the pre-estimate in the Appendix suggests, but if the original pre-estimate can be proved to be reasonable and realistic then the amount will stand. However, if it turned out that the pre-estimated rate and amount was totally disproportionate and was in fact a penalty upon the Contractor, the Courts could rule otherwise.

Although this clause gives the Employer the right to claim damages, he may not do so, either totally or in part, and thus exercise his discretion as built into the clause. Regardless of his decision, the Employer must give written notice to the Contractor intimating his intention before the issue of the Final Certificate.

After his written notice to deduct damages, they may be deducted from monies due in any Interim Certificate or the Final Certificate, and the written notice will adequately cover the Employer's obligation under Clause 30.1.1.3 to state the reasons for a deduction being made.

If the amount of damages exceeds the monies due in any certificate, the balance becomes a debt and can be recovered in the Courts.

When the Architect reviews his awards of extension of time within the 12-week period after Practical Completion, he may fix a later Completion date and thus cause an over-payment of damages to be made. The clause allows for payment to the Contractor in these circumstances, without any interest being added.

The position of the Nominated Sub-contractor is protected, as when a deduction is made from monies due and part of that is retention, any retention therein properly apportioned to a Nominated Sub-contractor cannot be used.

Finally, if the damages are stated in the Appendix as NIL and the Employer finds that he has been mistaken in this pre-estimate, he will most certainly have to stand by his mistake; it would be very difficult to raise an effective argument at a later date.

Clause 27 – Determination by the Employer

A very important distinction must be made in this clause, it is only the employment of the Contractor that is being determined and not the Contract itself. Therefore, with the Contract still valid, such matters as responsibility for the work executed, payments, settling accounts and of course arbitration, can all still operate. If the Contract was determined the aggrieved party could only sue for breach.

27.1   For the Architect to exercise his discretion as to determination, the Contractor must have defaulted in the first place and he must take certain specific actions. The specified reasons for default are

27.1.1   If without reasonable cause the Contractor wholly ceases the carrying out of the works before completion.

27.1.2   If the Contractor fails to proceed regularly and diligently with the works.

27.1.3   Refusal or persistent neglect to comply with a written notice from the Architect to remove defective work or improper materials or goods and that refusal and neglect materially affects the works.

27.1.4   Failure to comply with Clause 19 – Assignment and sub-contracts.
Local Authorities Edition only, failure to pay wages, etc. in accordance with the National Working Rule Agreement, Clause 19A.

The specific actions required by the Architect, and the time intervals between them, are set out graphically in *Diagram 12*.

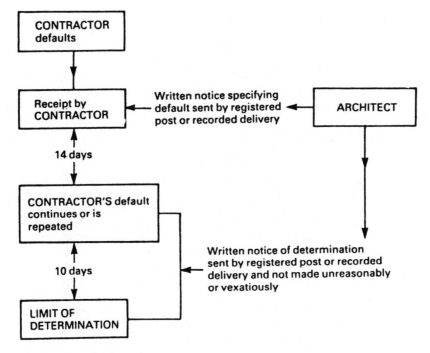

Diagram 12:

If the default ceases during the 14-day period, the determination will not take place since the notice from the Architect is no longer valid.

If the Architect does not issue his notice of determination by the end of the 24-day total period, again any determination notice would become invalid unless the Contractor defaults again and the procedure would begin anew.

27.2   Bankruptcy or liquidation is a default with ensuing automatic determination. No notice is required from the Architect but he would be advised to notify whosoever is appointed, trustee, receiver or liquidator, that the rights to determination are being invoked.

The clause gives the Employer and the trustee, receiver or liquidator the opportunity to come to some agreement that the Contractor's employment could be reinstated. This would mean that the person now operating the contractor's business takes over all his obligations under the Contract and also receives the rights of the Contractor under the Contract.

27.3   This clause applies only in the Local Authorities Edition and it gives the Employer the right to determine because of the Contractor's corrupt practices. The local authority does not need either to give notice or to specify the default.

27.4   If the employment of the Contractor is determined and he is not reinstated, then the detailed provisions which constitute the rights and obligations of both parties are as follows:

Employer   Can employ and pay others to carry out and complete the works and they can enter and use all temporary buildings, plant, tools, equipment, goods and materials and may buy all goods and materials necessary.

Comment   This can not apply to hired huts, plant, tools and equipment; the Employer will have to pay for their hire either before or after determination and this becomes costs incurred by the Employer to be settled at final account stage.

Contractor   Excepting determination caused by bankruptcy, etc., at the Employer's request within 14 days of the determination the Contractor will assign to the Employer, without payment, the benefit of any agreement for the supply of goods and materials or execution of any work, but subject to the Supplier or the Sub-contractor being entitled to object reasonably to any further assignment by the Employer.

Comment   This assignment can be accomplished only by a legal agreement to allow the provision of an objection to further assignment by the Employer being implemented. Further assignment can mean a substitute contractor.

Employer   Excepting determination being caused by bankruptcy, etc., the Employer can pay any supplier or sub-contractor, whether before or after the date of determination, unless payment has been made by the Contractor.

Comment   This is regarding domestic suppliers and sub-contractors and is payment of monies either outstanding at the date of determination or incurred after the date. It does not affect direct payments to nominated sub-contractors under the provisions of the Sub-contracts.

Contractor   When requested by the Architect in writing (but not before) he will remove off site any temporary buildings, plant, tools, equipment, goods and materials, either on hire or belonging to him.

Employer   If the Contractor does not comply as above, the Employer can remove and sell them and keep the proceeds to the Contractor's credit, less all costs. The Employer is not responsible for any loss or damage.

Comment   This removal could be requested immediately after determination or at completion of the works. If the Contractor still has a viable business, such amounts arising from the sale will be taken into account at final settlement.

   If, however, bankruptcy or liquidation causes the determination, the monies are paid to the person appointed to look after the Contractor's affairs and then will be available to all his creditors.

Contractor   Will allow or pay to the Employer the amount of any direct loss and/or damage caused to the Employer by the determination.

Employer   Is not bound to make any further payments to the Contractor until the works are completed by others. On completion and verification of the accounts within a reasonable time, the Architect will certify the amount of expenses and direct loss and/or damage due to the Employer.

If such amounts, when added to the monies already paid to the Contractor before determination, exceed the total amount which would have been payable at the completion of the Contract, the difference is a debt owed to the Employer,

OR

If such amounts, when added to the monies already paid to the Contractor before determination, are less than the total amount which would have been payable at the completion of the Contract, the difference is a debt owed to the Contractor.

Comment   Note that the Contractor is not paid until the end of the Contract, thus allowing for loss and expense exceeding amounts due to him.

Note also that 'direct loss and/or damage' is the phraseology here; therefore, there appears to be no right of the Employer to Liquidated and Ascertained Damages which might ensue after determination. Direct loss and/or damage is presumably then total and able to be proved in the Courts.

Determination caused by bankruptcy or liquidation will mean that the Employer's claim against the Contractor falls into the category of that of any other creditor.

The statement of a final account is to be ascertained by the Architect and would be made up as follows:

Amount of final account as if the Contractor had completed the works.
Less: Previous payments
Costs of the Employer completing the works (that is, all costs including professional fees)
Direct loss and/or damage caused to the Employer.

As this clause envisages amounts owing becoming debts, and as some of the sub-clauses could be contentious, it would be advisable to have legal advice on the whole matter of determination, if it occurs.

Clause 28 – Determination by the Contractor

This clause is the other side of the coin, where the Contractor can determine his own employment and, as in determination by the Employer, it is the employment and not the Contract itself that is determined.

The Contractor's option to determine his employment is based on grounds as follows (remember, it is on his own option, as no automatic determination takes place and it must not be unreasonable or vexatious):

Grounds

28.1.1   If the Employer does not pay the amount properly due in any certificate (except for a VAT adjustment) within 14 days from date of issue, and continues to default for seven days after receiving a registered post or recorded delivery preliminary notice from the Contractor, intimating his intention to determine within a further seven days from the date of the preliminary notice.

The sequence therefore is

(a) 14-day period for certificate to be honoured,
(b) After 14-day period, preliminary notice issued,
(c) Seven-day period for certificate to be honoured, starting on the date of the preliminary notice,
(d) After seven-day period has elapsed, notice of determination issued by registered post or recorded delivery.

The Contractor will have to be careful in his assessment of the amount properly due. The Employer can deduct from monies due under Clauses 30.1.1.2, 30.6.2.5 and 31.7, but the Contractor should have written information accompanying the certificate under these clauses.

28.1.2   If the Employer interferes with or obstructs the issue of any certificate.

It is clear that any certificate refers to more than just payment certificates, but only those with a direct financial relationship would apply, as under:

17.1   Practical Completion.
17.4   Completion of Making Good Defects.
18.1.1 Partial Practical Completion.
18.1.3 Partial Completion of Making Good Defects.
24.1   Failure to complete by Completion Date.
30.1   Interim Certificates.
30.8   Final Certificate.

This is a difficult matter to prove; has a certificate been deliberately under-valued, or held back, or any certificate not been issued at its proper time? The Contractor has not only to be satisfied that this has taken place, but also be satisfied that it was due to interference or obstruction by the Employer.

28.1.3   If the carrying out of the works still uncompleted is suspended for a continuous period of time named in the Appendix because of the list of events contained in Clauses 28.1.3.1 to 28.1.3.7.

The important points here are that the whole of the works has to be suspended and the period must be continuous. An aggregate of periods adding up to the stated period does not qualify. The periods usually stated in the Appendix are three months for loss or damage by Clause 22 Perils and one month for any other cause. Different periods could have been entered if desired, but the two periods must be stated in the Appendix.

Note that exceptions occur, namely:

17　　　　　Making good defects.

28.1.3.2　Loss or damage by Perils if is proved to be the Contractor's negligence.

28.1.3.4　Architect's Instructions if caused by Contractor's negligence or default.

28.1.3.7　Opening-up for inspection, if it is proved that materials and workmanship are not in accord with the Contract.

28.1.4　Bankruptcy or liquidation of the Employer gives grounds for determination by the Contractor, but only in the Private Edition. Presumably by omitting this clause from the Local Authorities Edition, the Employer cannot become bankrupt, but be careful who the client is when entering into a contract under the Local Authorities Edition.

Remember that under any ground for determination of a Contractor's employment, it is without prejudice to any other rights and remedies that he may have under the Contract or at Common Law.

Only Clause 28.1.1 has a set procedure to follow for determination. Under any other grounds a written notice is given, sent by registered post or recorded delivery, containing the reason or provision for determination and not being sent unreasonably or vexatiously.

28.2.1　When determination has taken place the Contractor will, as soon as possible, remove his temporary buildings, plant, tools, equipment and goods and materials from the site and let his sub-contractors do likewise.

Note that the Contractor is still liable for indemnification to the Employer against claims for personal injury or death, and damage to property (Clause 20) during the removal from site. Note also that goods and materials properly paid for by the Employer, are his property.

Lastly, until removal is complete, all rights and obligations of both parties are still operative; thus, any previously-agreed happenings, such as Liquidated and Ascertained Damages or direct loss and/or expense, can be either allowed or paid for as the case may be.

28.2.2　A settlement or final account has to be made to the Contractor, presumably by the Architect or the Quantity Surveyor since the Conditions do not delegate a person.

The final account will contain and be made up as follows:

28.2.2.1　Total value of work completed at the date of the determination made up as if it were an interim valuation under the rules of Clause 30.1.

28.2.2.2　Total value of work begun and executed but not completed, made up as if it were a variation under the rules of Clause 13.5.

28.2.2.3　Direct loss and/or expense ascertained under Clauses 26 and 34.3, whether before or after the date of determination.

28.2.2.4 Payment for materials and goods ordered for the works which the Contractor has paid for or is still legally bound to pay for (goods become the Employer's property after his payment).

28.2.2.5 The reasonable cost of removal of temporary buildings, etc. from the site.

28.2.2.6 Any direct loss and/or damage caused to the Contractor or any nominated sub-contractor due to the determination.

Less: Total amounts previously paid by the Employer.

28.2.3 The Employer must inform the Contractor, and any nominated sub-contractors, of amounts payable to nominated sub-contractors contained in the final account.

As a postscript to this clause, bankruptcy on the part of the Employer is a ground for determination. This would make the Contractor just another creditor and cause him to claim against the person appointed to take over the Employer's business.

Note that the balance due under this clause is not described as a debt to either party.

Scotland

Clause 24 – Damages for Non-completion

Clause 27 – Determination by the Employer

Clause 28 – Determination by the Contractor

Clause 24 – Damages for Non-completion

There are no changes to this clause for use in Scotland.

Clause 27 – Determination by the Employer

The Scottish Supplement to the Scottish Building Contract deletes Clause 27.2 in whole and substitutes three sub-clauses.

27.2.1 This clause grants discretion to the Employer either to determine or to await the findings of a provisional liquidator before determining. The provisional liquidator has the right to seven days' notice from the Employer before the determination notice is issued. This is different from England and Wales where the determination is automatic.

27.2.2 If the Contractor becomes bankrupt or goes into liquidation, then determination is automatic, but courtesy alone might dictate an intimation with regard to determination being made to the trustee or liquidator.

27.2.3 Again, discretion is granted to the Employer to allow him to enter into an agreement with the trustee or liquidator to reinstate the Contractor's employment. In such a case it is the trustee or liquidator who accepts the liabilities of the Contractor and also his rights, under the Contract.

Clause 27.4.3 is changed to allow the Employer to 'sell any such property as far as belonging to the Contractor'. This clarifies the position regarding property on hire or under a hire-purchase agreement, not being included.

A very important difference in Scotland is that immediate arbitration is available over disputed determination, as stated in Clause 41.1.4 of the Scottish Supplement and in the Scottish Building Contract.

Assignation under Clause 27.4 gives the Employer the entitlement to have assigned to him such sub-contract or supply agreements as he requires. The involvement here of Nominated Sub-contractors and Nominated Suppliers is made clear by the respective additional Clauses 35.27 and 36.6 in the Scottish Supplement.

Clause 28 – Determination by the Contractor

Only Clause 28.1.4 is altered in the Scottish Supplement and this is an insertion to bring the wording into line with Scottish law and practice.

Again, immediate arbitration is available over disputed determination, as stated in Clause 41.1.4 of the Scottish Supplement and in the Scottish Building Contract.

Chapter 16

Other matters

Clause 32 – Outbreak of hostilities

Clause 33 – War damage

Clause 34 – Antiquities

England and Wales

Clause 32 – Outbreak of hostilities

32.1   During the currency of the Contract, if there is an outbreak of hostilities, either party has the option to determine the employment of the Contractor. However, there are provisions, as follows:

  (a) Hostilities must be of a major kind.
  (b) The United Kingdom must be involved, but not be necessarily the Theatre of War.
  (c) General mobilization must have taken place.

If all the above provisions are satisfied, a notice of determination may be sent by and to either party, as follows:

  (a) The notice to be in writing and sent by registered post or recorded delivery.
  (b) Before the notice is served, a time lapse of 28 days, starting from the date of hostilities, must be observed.
  (c) The notice cannot be served after Practical Completion unless the works, or any part of it, has sustained war damage.

The war damage to the works mentioned in (c) above ensures that the Contractor retains his obligations to make good defects, but not to make good the war damage.

32.2   Determination takes place immediately on receipt of a notice from either party, and within 14 days of that date the Architect can issue instructions to the Contractor for protective work to be carried out, or for the works to be continued to some convenient stopping point. The Contractor must comply with these instructions as if determination had not taken place.

The Contractor has a time limit of three months in which to complete the work in the instructions, and if by that time he cannot complete, for reasons outside his control, he can abandon the work.

32.3   A final account is prepared by the Architect all in accordance with Clause 28.2 – Determination by the Contractor, but without the right in Clause 28.2.2.6 to claim direct loss and expense. The only addition to the account not allowed for in Clause 28.2, is payment for the work in the Architect's Instructions, if any, which will be valued as if they were variations.

This final account cannot of course be ascertained or a payment made until 14 days after the date of determination, or the completion of the work in the Architect's Instructions, or abandonment, whichever is the latest.

There is, as in Clause 28.2, no fixed period or date for the final payment to be made, only commencement of the valuation.

The footnote to Clause 32 allows both parties to ignore the rules of procedure in the clause, if they wish to come to some private agreement to meet the circumstances arising from an outbreak of hostilities.

Clause 33 – War damage

This clause begins back-to-front, as the definition of war damage is the last Sub-clause 33.4, which merely gives the War Damage Act 1943 as the authority for the definition. If war damage did in fact happen, that Act would need scrutiny to confirm not only that the happening to the works came within that definition but indeed further that the Act is still current.

The clause is really in two parts, the first part referring to the procedures to be followed without the complications of determination.

33.1   If war damage takes place, this clause has precedence and cancels all other clauses in the Conditions, and the following measures will be taken:

(a) Any amounts of completed work to be valued shall be ascertained regardless of war damage. Even if the building is destroyed, the Contractor must be paid for his work.

(b) The Architect can issue instructions regarding removal and disposal of debris, or damaged work, and to erection of any protective work required.

(c) The Contractor to reinstate, make good war damage, and carry on and complete the works.

(d) The Architect to give a notice in writing fixing a fair and reasonable new completion date. This is an automatic extension of time.

(e) Payment for removal and disposal, protective work and reinstatement will be under the rules of payment for variations.

The second part of the clause deals with the possibility of determination becoming involved and raises an interaction with Clause 32.

33.2   This clause envisages the situation where war damage was the cause of determination in Clause 32, or it happened after a determination notice was issued. In these circumstances, all Architect's Instructions given under this clause are deemed to have been given as if under Clause 32 and the procedures and final account are prepared in accordance with the rules of that clause. In essence, Clause 33 ceases and Clause 32 takes over.

33.3   The Employer is entitled to payment from the Government in respect of war damage, all in accordance with any prevailing regulations. This payment does not have any effect upon the Contractor's entitlements to payment under the clause, either in the amount or when he is paid.

Note here that the Contractor will not be paid for loss or damage to his own plant, etc., nor materials and goods offsite.

Clause 34 – Antiquities

Generally, this clause relates to any objects previously unknown which may be discovered during operations. The phrase 'of interest or value' could cover almost anything. The Contractor should be cautious and follow the rules here, rather than make up his own mind as to the importance of any object. All objects become the Employer's property which, in turn, might be claimed for treasure trove.

The clause is also in being to prevent objects from being damaged through inexpert attention, and takes into account that the position may be important. Sadly, the first persons to encounter such an object may be a labourer or machine operator, who may not be interested. Perhaps standing instructions regarding strange objects might not go amiss.

The quickest way to report such a find is to the Clerk of Works, and this should be done immediately.

Experts can be brought to the site and they will be classed broadly the same as those in Clause 29, not sub-contractors and not the Contractor's liability to indemnify from Clause 20 insurances.

34.1   If the Contractor discovers any such object of interest or value, which it is not his liability to assess, he will

    (a) Immediately inform either the Clerk of Works or the Architect of the discovery and give its precise location.

    (b) Not disturb the object, ceasing work if necessary.

    (c) Take all precautions to preserve the object in its exact position, shoring, temporary fencing or whatever is necessary.

    (d) Give written notice to the Architect by registered post or recorded delivery, stating all the relevant facts and requesting his instructions. The clause does not require this notice to be given, but it is necessary in the Contractor's own interests. An extension of time will not start until after the Architect has issued instructions; if he does not issue them the Contractor's notice will form a start point, and he, of course, will be anxious to keep any waiting time between discovery and an instruction to a minimum since this period does not count towards an extension of time.

34.2   The Architect must issue an instruction to the Contractor regarding the discovery and the latter's actions as regards the works. Here the Architect's powers are very wide, deliberately so, as there can be no preconception of how such a discovery could affect the works.

34.3   As the issue of the Architect's Instruction makes the discovery a Relevant Event, extension of time can be granted by the Architect. Therefore, after receipt of the instruction the normal procedures should be gone through as in Clause 25 for any other extension of time claim.

If the Architect issues a postponement of the works which causes the period stated in the Appendix to be over-run, there is no case for determination of the Contractor's own employment under Clause 28.

Determination may be possible under this clause only if the Architect did not issue an instruction after receipt of the Contractor's notice requesting one. This would depend upon the over-run of the period in the Appendix, a request from the Contractor for an instruction, and non-compliance by the Architect. As such a discovery would most probably occur at the beginning of a contract, possibly during the excavations, such a course as determination would be a very drastic step.

The Architect can also reimburse any direct loss and/or expense caused to the Contractor by the discovery and its consequences. The clause makes this an automatic reimbursement although it is in the Architect's opinion. No formal request is needed from the Contractor, but again in his own interests it would be wise to go through the Clause 26 procedures.

The Contractor will have to ensure that the reimbursement is adequate since there is no breach or default on the Employer's part, thus preventing the Contractor from bringing an action for breach.

The reimbursement covers compliance with the Architect's instructions and also loss and expense caused to the Contractor by his own actions or the cessation of action.

As the procedure in Clause 26, any portion or all of the ascertained amounts are added to the Contract Sum and paid in Interim Certificates without retention.

Scotland

Clause 32 – Outbreak of hostilities

Clause 33 – War damage

Clause 34 – Antiquities

There are no changes to any of the above clauses for use in Scotland.

It should be noted here, however, that both Clause 32 and Clause 33 have immediate arbitration available over disputes as stated in Clause 41.1.4 of the Scottish Supplement and in the Scottish Building Contract.

Chapter 17

Fluctuations

Clauses 37 to 40 – Fluctuations

England and Wales

Clause 37

This clause indentifies the three alternatives which are available for recovering fluctuations, namely

Clause 38 – Contributions, levy and tax

Clause 39 – Labour, materials, contribution, levy and tax

Clause 40 – Price adjustment formulae

The three clauses above are published separately from the Conditions and incorporated therein by reference, the documents being issued by JCT and listed as under:

Fluctuation clauses for use with the Private Editions with, without and with Approximate Quantities.
Fluctuation clauses for use with the Local Authorities Editions with, without and with Approximate Quantities.

There is also a separate document for Nominated Sub-contracts NSC/4 and NSC/4a.

The Appendix has an entry in which the three clauses are listed, with the two inapplicable clauses having to be struck out. If no clause is struck out, then Clause 38 is deemed to be in operation.

The three clauses can be defined or described as

Clause 38 to be used only for the recovery of contributions, levy and tax fluctuations which are amounts payable by a person in his capacity as an employer – in this case the Contractor, since he employs labour. This is commonly known as 'fixed price'.

Clause 39 to be used where the parties have agreed to allow labour, material and tax fluctuations to be recovered. This is sometimes known as 'full fluctuations'.

Clause 40 to be used where the parties have agreed to fluctuations being dealt with by adjustment of the Contract Sum under the Price Adjustment Formulae for Building Contracts. Mostly known as the 'Formula Method' and is also 'full fluctuations'.

Clause 38 – Fixed price

This clause provides only for an adjustment of the Contract Sum in regard to changes in government or statutory levy and tax contributions paid by an Employer. It also affects import duties on materials and goods and any other restrictions on purchase or sale.

These fluctuations are minimal and do not occur with any great frequency, thus allowing the Contractor a reasonable opportunity to forecast the possible effect in advance. Thus most Quantity Surveyors advise that this clause be used in the case where a contract will not exceed 12 months' duration. Of course, this will also mean the Contractor having to forecast any possible increase in labour and material costs during that period, as by using this clause recovery of fluctuations pertaining to them is excluded and has to be allowed for in the general pricing levels.

38.1   For the purpose of this clause the prices in the Bills of Quantities are deemed to have been based on types and rates of contributions, levy and tax fixed by Act of Parliament and current at date of tender. The exception here is the Construction Industry Training Board levy which is not levied under an Act of Parliament. The date of tender is defined in Sub-clause 38.6.1.

38.1.2   If any new tax or levy is introduced or an existing one is increased, then the net difference payable by the Contractor in respect of his workpeople is reimbursed to him.

Similarly, if a tax or levy is removed or an existing one decreased, then the net difference payable by the Contractor in respect of his workpeople is allowed by him.

'Workpeople' in this context means not only those persons working directly on the site but also those adjacent to it. Also covered are persons who are producing goods and materials for future incorporation in the works but do not need to be working either on site or adjacent to it. This of course will include Domestic Sub-contractors.

38.1.3   Site staff too are included in fluctuations and become eligible for reimbursement or allowance by the Contractor as the case may be. These site staff are outwith the definition of workpeople and represent all staff employed on site.

38.1.4   The clause sets down rules for the elegibility of the site staff to be included. The minimum time on site in any working week must be two whole working days, and parts of days cannot be aggregated to arrive at the minimum.

The rate for calculating the amount payable or allowable to or by the contractor in respect of site staff will be at the highest trade operative's rate paid by the Contractor or a Domestic Sub-contractor.

Under the provisions of Clause 38.4.5 a weekly return by the Contractor is required certifying his claim for fluctuations.

38.1.5 to **38.1.7**   These clauses specifically refer to any premiums or refunds to the Contractor in consequence of his employment of labour. This might refer to premiums or subsidies, or where any tax might be refunded. The rules here are exactly the same as for Clause 38.1.2.

38.1.8   'Contracted-out' workpeople are those voluntarily not contributing to part of the State pension scheme by reason of some other private occupational-type pension scheme of which they are members.

The clause makes it clear that fluctuations are based on the Employer's contributions to the State scheme regardless of the status of the workpeople.

38.1.9   This is the definition of a contribution, levy and tax.

38.2   Here similar rules are provided for changes caused by Acts of Parliament for the undermentioned, on the import, purchase, sale, appropriation, processing, or use of

(a) Materials, including temporary formwork,
(b) Goods, including temporary formwork,
(c) Electricity,
(d) Fuels.

With regard to fuels, it will have been made clear in the Tender documents whether or not this is an allowable item. Among items (a) to (d) only those listed by the Contractor on a page attached to the Contract Bills will qualify.

VAT is specifically omitted from the calculations.

38.3   If the Main Contract Conditions in regard to fluctuations are regulated by Clause 38, the Domestic Sub-contractors must reflect these same conditions in order that their fluctuations will be treated in a like manner.

Fluctuations are paid or allowed to or by the Main Contractor for the Domestic Sub-contractors and he in turn adjusts the Sub-contracts.

38.4.1 to **38.4.6**   The following procedures have to be observed:

(1) Contractor gives a written notice to the Architect when any fluctuation occurs in the listed clauses.
(2) The written notice to be given within a reasonable time, and it is a pre-condition for payment to be made.
(3) The Quantity Surveyor and the Contractor will agree the net amount of the fluctuations.
(4) Number (3) is dependent upon the Contractor submitting documentary evidence as required to enable the amount to be computed, both on his own behalf and of that of his domestic sub-contractors. This information to include a weekly certification of the evidence.
(5) The net amount to be added to or deducted from the Contract Sum, paid by inclusion in Interim Certificates and not subject to any profit that the Contractor has included in his Contract Sum.

38.4.7   This clause is sometimes referred to as the 'freezing conditions' which, put simply, means that no adjustment for fluctuations will be made after the Completion Date (as stated in the Appendix) while the works have not reached Practical Completion.

38.4.8   For the above clause to operate, two conditions must be fulfilled:

(a) Clause 25 (extension of time) is unamended and not deleted from the Conditions.
(b) The Architect has properly fixed and confirmed in writing any new Completion Date in response to all written notices from the Contractor under Clause 25.

38.5   Fluctuations do not apply in this clause to the following:

(a) Daywork,
(b) Work by nominated sub-contractors and suppliers,
(c) Work by the Contractor on his own Tender under Clause 35.2,
(d) Value added tax.

38.6   This is a list of definitions.

38.7   The Appendix has a stated percentage to be added to the net amount paid to or allowed by the Contractor, which is entered by the Employer. The percentage is to make some allowance for other members of the Contractor's staff not included in the clause, such as those in head office.

Clause 39 – Full fluctuations

This clause provides for an adjustment of the Contract Sum in regard to labour, materials and levy and tax contributions paid by an employer.

Most Quantity Surveyors would advise the use of this clause or that of Clause 40 where a contract will exceed 12 months' duration.

The clause can be divided under several headings, thus:

Labour,
Transport charges and fares,
Contributions levy and tax,
Materials, goods, electricity and fuels,
Domestic sub-contractors.

Labour

39.1   For the purposes of this clause the prices in the Bills of Quantities are deemed to have been based on rates of wages, additional emoluments and expenses including holiday credits, payable by the Contractor to his workpeople.

'Workpeople' are defined again here as those employed directly on site, together with those working adjacent to it, and other persons who are producing goods and materials for future incorporation in the works. This includes domestic sub-contractors.

The wages paid must be in accordance with

(a) The National Joint Council for the Building Industry or other wage-fixing body which is applicable.

(b) These wage rates have to be promugated at the date of tender ('promulgation' means publication date of the wage rates).

(c) Any incentive scheme or productivity agreement under the rules of the National Joint Council for the Building Industry (unofficial schemes do not qualify).

(d) The Terms of the Building and Civil Engineering Annual and Public Holiday Agreements, or other similar wage-fixing body.

(e) Any amounts of contributions, levy or tax which the Contractor has to pay in his capacity as an employer, calculated by reference to rates of wages, other emoluments and expenses.

39.1.2   If any rates of wages or other emoluments and expenses, are increased or decreased after date of tender, the following net amounts are paid to or allowed by the Contractor:

(a) The net difference in cost of wages,

(b) The net difference in cost of other emoluments and expenses (this refers to bonus, overtime, tool money, dirty money, etc.),

(c) The net difference in cost of holiday credits,

(d) The net consequential difference in

Employer's liability insurance,
Third party insurance,
Contribution, levy or tax based on wages.

39.1.3 to   The addition of site staff being included in labour fluctuations and the
39.1.4   rules as to their eligibility is exactly the same as set out in Clause 38, therefore the comments on Clauses 38.1.3 and 38.1.4 are relevant here.

Transport charges and fares

39.1.5   For the purposes of this clause the prices in the Bills of Quantities are deemed to be based upon transport charges to transport workpeople or on fares paid to workpeople under the Working Rule Agreement promulgated at Date of Tender, or some other wage-fixing body.

Note that the Contractor has to submit a list of basic transport charges which is attached to the Contract Bills, if this is the method which he uses to price his Bills.

The net difference in cost of transport charges between the basic list and any charges after date of tender will be paid to or allowed by the Contractor.

OR

The net difference in the cost of fares.

Note here that site staff are excluded.

Contribution, levy and tax

39.2 to   These clauses are exactly the same as Clauses 38.1.1 to Clause 38.1.9
39.2.8   as set out in Clause 38, and the comments on those clauses are relevant here.

Materials, goods, electricity and fuels

39.3   For the purposes of this clause the prices in the Bills of Quantities are
deemed to be based upon the market prices of the materials, goods,
electricity and fuels.

A list of basic prices of the above, to be attached by the Contractor
to the Contract Bills, forms an exclusive list of materials, goods,
electricity and fuels on which the Contractor wishes fluctuations to
operate. Omissions from the list will not be considered. The basic list
prices are current at date of tender.

Again, fuels will have been stated in the Tender Documents as to
whether or not they can be included.

Under the definitions in Clause 39.7.2 'goods and materials' include
timber used in formwork but do not include other consumable stores,
plant and machinery. Electricity would require to be listed as a price
per unit with it being made clear which phase the supply would be in.
'Fuels' (if allowed) referes to all fuels – petrol, diesel, oil, grease, etc.

Any fluctuations will be the net difference in cost between the
materials, goods, electricity and fuels listed by the Contractor and the
market prices after date of tender.

Note that the electricity and fuels used in carrying out the
Contractor's temporary works is also included. Any changes in VAT
are excluded.

Domestic sub-contractors

39.4   Again, under this clause the Contractor is obliged to ensure that
matching fluctuation arrangements are included in all his domestic
sub-contracts. As before, the Main Contractor recovers all fluctuations
on behalf of his domestic sub-contractors and then adjusts the
sub-contracts. If the fluctuation arrangements are not included in the
domestic sub-contracts, no recovery of these fluctuations will operate
and technically the contractor is then in breach of contract with the
Employer, as this clause is an obligation upon him.

39.5.1 to   The following procedures have to be observed:
39.5.6
(1) Contractor gives a written notice to the Architect when any
fluctuation occurs in the listed clauses.
(2) The written notice to be given within a reasonable time and is a
pre-condition for payment to be made.
(3) The Quantity Surveyor and the Contractor will agree the net
amount of the fluctuations.
(4) Number (3) is dependent upon the Contractor submitting
documentary evidence as required to enable the amount to be
computed, both on his own behalf and on that of his domestic
sub-contractors. This information to include a weekly certification
of the evidence.
(5) The net amount to be added to or deducted from the Contract
Sum, paid by inclusion in Interim Certificates and not subject to
any profit that the Contractor has included in his Contract Sum.

39.5.7   Once again this is the clause sometimes referred to as the 'freezing conditions', meaning that no adjustment for fluctuations will be made beyond the Completion Date stated in the Appendix, while the works have not reached Practical Completion.

39.5.8   As before, two conditions have to be fulfilled before the above clause can operate:

(a) Clause 25 (extension of time) is unamended and not deleted from the Conditions.
(b) The Architect has properly fixed and confirmed in writing any new Completion Date in response to all written notices from the Contractor under Clause 25.

39.6   Fluctuations do not apply in this clause to the following:

(a) Daywork,
(b) Work by nominated sub-contractors and suppliers,
(c) Work by the Contractor on his own Tender under Clause 35.2,
(d) Value added tax.

39.7   This is the list of definitions.

39.8   The Appendix has a stated percentage to be added to the net amount paid to or allowed by the Contractor, which is entered by the Employer. The percentage is to make some allowance for other members of the Contractor's staff not included in the clause, such as those in head office.

Clause 40 – Formula method

The essential difference between this clause and the two preceding clauses is that whereas under Clauses 38 and 39 the actual physical increases in costs are claimed by the Contractor and recovered from the Employer, under Clause 40 the fluctuation amounts are a calculation based on pre-determined rules, thus being theoretical amounts as opposed to actual amounts.

40.1   The formula method has 48 work categories which include all work of a similar nature within each category. Contractor's specialist work is where there is no work category, but for which one of the specialist engineering formulae is appropriate. Where no work category exists for certain work to be allocated under, it is totalled and put under the heading of 'Balance of Adjustable Work'. This will include preliminary and general items, water, insurance, general and special attendance upon nominated sub-contractors, adjustment of the tender summary and similar matters. This balance of adjustable work is calculated as a proportion of the amount allocated to work categories.

The Contract Sum will be adjusted in accordance with Clause 40 and the Formula Rules current at date of tender. VAT will not be included or considered.

Rule 3 of the Formula Rules defines the terms used.

All adjustments made under the Formula Rules will be included in Interim Certificates and any errors corrected in further certificates.

Rule 5 of the Formula Rules covers the errors which shall be corrected:

5a (a) Arithmetical errors,
  (b) Incorrect allocation of value of work categories,
  (c) Incorrect allocation of work as Contractor's specialist work,
  (d) Use of an incorrect index number.

5b If the Contractor questions matters referred to in 5a above in relation to any valuation of a certificate, the Quantity Surveyor will make available all relevant documents to allow the Contractor to make his own inspection, provided that the request is made within a reasonable time after the certificate.

40.2   This clause makes it obligatory for the Quantity Surveyor to prepare a valuation for fluctuations before each Interim Certificate, and thus amends Clause 30.1.2.

40.3   This clause applies to articles and goods manufactured outside the United Kingdom and imported by the Contractor for use in the works. As the indices are based upon goods and materials of British make, these imported articles and goods are dealt with separately.

At tender stage the Contractor will append to the Contract Bills a list of foreign-made articles, stating also their market price. If, after date of tender, the market price varies, the net difference will be paid to, or allowed by, the Contractor. 'Market price' also refers to any taxes imposed, such as an import tax, but excludes VAT.

40.4   In regard to nominated sub-contractors, their separate sub-contract will determine how the fluctuations are to operate, which is included in the Tender of the Nominated Sub-contractor and accepted by the Architect.

Domestic sub-contractors can have the same Clause 40 fluctuations agreement if so desired, and matching provisions can be made to the Main Contract. However, unlike Clauses 38 and 39, amounts paid to the Main Contractor for fluctuations cannot be applied directly to his domestic sub-contractors. Therefore, any method of regulating fluctuations under Clause 40 could be incorporated in the domestic sub-contract.

Note that it is not obligatory here to mirror the Main Contract fluctuations if the Contractor and the Domestic Sub-contractor agree to some other arrangement.

40.5   This clause allows the Contractor and the Quantity Surveyor to agree to some other method or procedure of the Formula Rules. However, two conditions must operate:

(a) The amount ascertained of formula adjustment has to be the same, or approximately the same, as if the method had not changed.
(b) The amount to be paid to any domestic or nominated sub-contractor will not be affected.

In the context of (b) above, the Nominated Sub-contracts also allow such an agreement to be made, which could accord with the Main Contract agreement, if so desired.

40.6   Indices are published monthly and this clause envisages the need for an alternative procedure should they either fail to be published in time or miss a month altogether. The steps to be taken are as follows:

(a) A fair and reasonable adjustment to be made for that month.
(b) When or if re-publication takes place, the fair and reasonable amount will be re-ascertained and substituted by the proper amount.
(c) Records will be kept to enable any period of delay to be readily ascertained after commencement of the publication of the monthly bulletins.

40.7   Again, as in the other two clauses 'freezing conditions' apply, meaning that no adjustment for fluctuations will be made beyond the Completion Date, stated in the Appendix, while the works have not reached Practical Completion.

As before, two conditions have to be fulfilled before the above clause can operate:

(a) Clause 25 (extension of time) is unamended and not deleted from the Conditions.
(b) The Architect has properly fixed and confirmed in writing any new Completion Date in response to all written notices from the Contractor under Clause 25.

Note that this means that formula adjustment included in Interim and Final Certificates can only be made with the indices appropriate to the month which contains the Completion Date.

Clause 40 does not contain any percentage addition to the ascertained amounts, as the fluctuations are based on work categories obtained from the Bills of Quantities. Presumeably the pricing of the Bills of Quantities covers for overheads, site staff and Head Office Staff and overheads.

The Local Authorities Editions contain a clause to enable a deduction to be made for the non-adjustable element. This limit has been set at 10 per cent, which means that the total increased costs payable or allowable, are reduced by 10 per cent or paid at 90 per cent of their value. The reasoning here is that the Contractor should not be paid for increased costs on his overhead expenses.

Scotland

Clauses 37 to 40 – Fluctuations

There are no changes to the above clauses for use in Scotland.

Chapter 18

Additional clauses and information

Clause 15 – Value added tax – supplemental provisions
Supplemental Provisions (the VAT Agreement)
Clause 19A – Fair Wages
Arbitration
Differences between Local Authorities and Private Editions
England and Wales
Clause 15 – Value added tax – supplemental provisions

This clause incorporates the VAT Agreement into the Conditions only by reference in Clause 15. The VAT Agreement is printed separately after the Appendix to separate it from the Conditions. VAT does not form part of the Architect's valuation and certification obligations and he does not have to deal with any payment of VAT to the Contractor.

Value added tax operates under the auspices of HM Customs and Excise who are the VAT Commissioners. Any goods and services which are *purchased* may have VAT added to them (the 'input tax'). Goods and services which are *supplied* may have VAT added to them (the 'output tax'). Any business then, has to pay to HM Customs and Excise the balance left over when the input taxes are deducted from the output taxes. As the building industry has a zero rate chargeable for VAT on new buildings, but alterations and extensions to buildings are taxed, the balance left represents the total of the input tax, which is refunded to the Contractor on his making a proper claim to the commissioners.

15.1   This is purely the definitions relating to VAT.

15.2   Here it is made very clear that the Contract will be exclusive of VAT, but provides than any VAT which might become a charge to the Employer from the Contractor will be paid under the rules of this clause and the VAT Agreement.

15.3   This is a saving clause on the Contractor's behalf. It envisages the possible exemption or change in the tax arrangements occuring after date of tender. If exemption from tax came into being, the Contractor would be unable to offset his input tax, and this clause allows for the Employer refunding these imput taxes that he can no longer offset. This would mean that to allow this clause to operate the zero rating on the supply of goods and services in the building industry would have to be changed to 'exempt'.

The VAT Agreement – supplemental provisions

The following is the VAT Agreement referred to in Clause 15.1 and incorporated therein by that reference.

Clause 1   This clause sets out in principle that the Employer will pay to the Contractor any VAT that under the contract is properly chargeable for goods and services.

It further specifies the regulations under which Clause 1.4 will operate, Authenticated Receipts.

Clause 1.1   The Contractor is required to give the Employer his provisional assessment of the amounts included in each Interim Certificate, which are subject to VAT.

To enable the Contractor even to attempt this exercise, obviously he must have the valuation of the Interim Certificate before it is passed to the Architect. Therefore a close liaison with the Quantity Surveyor is needed to establish the amounts in time. Remember that the amounts in the assessment are only provisional.

The Contractor's duties here are as follows:

(a)   Specify the amounts less any retention.
(b)   State the tax rating, zero or otherwise, for each amount.
(c)   State the grounds upon which the tax rate is chargeable.
(d)   Submit all the information before the date of the Interim Certificate.

Note that under Clause 1.3 a final reconciliation of all the VAT is made. This procedure being carried out forms an exclusion to the Contractor's obligations under Clause 1.1

Clause 1.2   Unless the Employer objects to the provisional assessments, he has three working days in which to so notify the Contractor in writing. The Contractor in turn, upon receipt of the Employer's notification, has another three working days in which either to withdraw his assessments or to confirm them. If the Contractor does confirm in his reply to the Employer, then he can treat any amount received from the Employer in the Interim Certificate as being inclusive of a rated tax. He then issues an authenticated receipt based on his provisional confirmed assessment.

Consider the time scale involved here. The original provisional assessment to be in the Employer's hands prior to the issue of the Interim Certificate. This can only be immediately prior to the date of issue, since the Quantity Surveyor will pass on his valuation as soon as possible after its computation.

If both parties react with unusual alacrity, the position will be as shown in *Diagram 13*.

Payment of a Certificate has to be made within 14 days of the date of issue, and the time allowance for the two notices are each three working days. If weekends, public holidays or the transmission by post intervene, then the allotted time scale might be impossible. Notice that each party has three working days within which to make his reply, not including postage.

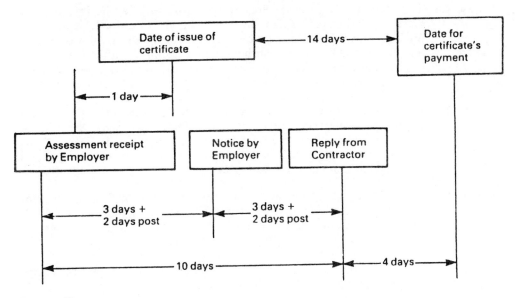

Diagram 13:

It may be that some arrangement can be made with the Employer prior to the start of the Contract, to enable some reasonable amounts to be allowed for each Certificate and a final assessment made under Clause 1.3 or the procedures under Clause 3.

Clause 1.3 This is the final reconciliation of the total VAT payable under the Contract. It is carried out by the Contractor as soon as possible after issue of the Certificate of Making Good Defects. His duty is to send a written final statement to the Employer containing

(a) The various values of all supplies of goods and services included in certificates already issued and still to be issued,

(b) State the rates of tax chargeable against each of these values,

(c) State the grounds upon which each tax rate is based, and

(d) State the total amounts of tax already received and for which receipts have been issued.

Note that the Contractor may wait until after the issue of the Final Certificate before issuing his statement and there would appear to be no time limit set beyond that date.

If the Employer is in agreement with the final statement, he uses the values and rates of tax to calculate the VAT due to the Contractor. The balance, if any, due to the Contractor, to be paid by the Employer within 28 days of the receipt of the statement. If, on the other hand, an overpayment of VAT has been made, the Employer will inform the Contractor and receive back the excess amount within 28 days of the Employer informing the Contractor of the excess. The Contractor will also include a receipt showing the adjustment of the amounts.

Clause 1.4   The receipts referred to here are special receipts authorised by HM Customs and Excise under the VAT Regulations. Each payment of VAT by the Employer must be properly receipted by the Contractor and these are known as 'authenticated receipts'. This receipt has a double purpose: it acts in the original for the Employer to justify his input tax, and as a copy for the Contractor to justify his output tax.

Clause 2   It is made very clear in this clause that the amount of VAT in a certificate cannot be used to deduct Liquidated and Ascertained Damages from, which might be due to the Employer. It also confirms that the final statement will take no account of Liquidated and Ascertained Damages.

Clause 3   The procedures set out here cover for the Employer's disagreement with the final statement as sent by the Contractor under Clause 1.3. The disagreement can be over the amounts, rates of tax, or grounds for the rates charged, and any part in disagreement is sufficient for this clause to come into effect.

The employer must inform the Contractor within 28 days of receipt of the final statement that he disagrees. Request the Contractor to obtain a ruling from HM Customs and Excise as to the proper amount of tax payable to the Contractor, which has already been paid by the Employer.

If the ruling so obtained still does not satisfy the Employer, he can request the Contractor to appeal to the VAT Tribunal for a binding decision. However, several conditions as follows must be fulfilled when this action is taken:

(a)   The Contractor must have paid to HM Customs and Excise the full amount of his assessment of the VAT.
(b)   The Employer must have paid to the Contractor the full amount of the assessment of the VAT.
(c)   The Employer must indemnify the Contractor against all costs and expenses of such an action.
(d)   The Employer, at the option of the Contractor, secures such costs and expenses (this could be a deposit in a bank account).
(e)   The Contractor will account for any costs awarded in his favour.

On one of the two dates, depending upon the action taken by the Employer, as follows:

(a)   28 days after the ruling of the Commissioners if no further appeal is made,

OR

(b) 28 days after the decision of an appeal to the VAT tribunal,

the Employer or the Contractor will pay or refund to the other in accordance with the Commissioners' findings, or the VAT Tribunal's findings, any over- or under-payment of tax. An Authenticated Receipt will also be issued to show the adjustment.

Clause 4   This clause simply states that payments of balances of VAT due to the Contractor, whether arrived at by a ruling or an appeal decision, will be final and the Employer's obligation fulfilled.

However, HM Customs and Excise have the right to make any correction to the amounts at any time after the final balance is paid. If these corrections again alter the amounts to be paid or refunded, this will be done by either party as necessary, and presumeably an Authenticated Receipt issued accordingly.

Clause 5   If any arbitration award alters the amount of any work which is subject to VAT, then the VAT payable must be altered in turn. This would mean a payment or refund and another Authenticated Receipt.

Clause 6   This clause makes it clear that arbitration cannot relate to the VAT Agreement itself, as it has its own built-in appeal procedure.

Clause 7   If the Contractor defaults in providing an Authenticated Receipt, the Employer can suspend all further payments to the Contractor. As the clause states, this is *all* payments, Interim Certificates and VAT payments. However, the following two conditions must apply before such a drastic step is permitted:

(a) The Employer must prove that his requirement for the receipt is to allow him to justify any claim for tax paid which he is making to the Commissioners.
(b) The Employer has paid the tax in accordance with the provisional assessment, unless a reasonable objection has been upheld.

Clause 8   This final clause allows for determination of the Contractor's employment by the Employer (Clause 27) having taken place. In that case it is possible that the second Contractor who finishes the work will generate more VAT amounts to be paid by the Employer. The clause allows the Employer to recover any additional VAT payment which is a direct cause of the determination, from the original Contractor as part of his over-all claim.

NB: Due to VAT changes under the April 1984 Budget, these provisions may be altered later.

Clause 19A – Fair wages

This clause appears only in the Local Authorities Editions.

It is an obligation upon the Contractor that he pays rates of wages and observes hours and conditions of labour at least equal to those of the National Joint Council for the Building Industry, or other wage-fixing body. The Contractor also warrants that he has complied with Clause 19A for at least three months prior to the date of his Tender.

The workpeople to be free to join trade unions.

Clause 19A, together with negotiated wage rates, hours and conditions of work, to be displayed in all places where his workpeople are engaged.

The Contractor to ensure that his Sub-contractors also abide by the general conditions of Clause 19A.

The Contractor to keep records establishing time worked and wages paid, and to produce them to the Employer, if required.

The Architect/Supervising Officer can request proof of payment of wages, rates of wages and hours and conditions observed by the Contractor and his Sub-contractors.

It would appear that one recourse for non-observance of the clause is by referral through the Minister of Labour to an independant tribunal.

Note, however, that non-observance is a ground for determination by the Employer under Clause 27.1.4.

Arbitration

Article 5 of the Articles of Agreement forming part of the contract relates to arbitration.

Article 5.1 Disputes and differences between the Employer and the Contractor and between the Architect and the Contractor (when the Architect is acting on the Employer's behalf) can be referred to arbitration. These referrals can be made during the progress of the works, after completion and up to 14 days beyond the date of the final Certificate, or after abandonment of the works. Either party can instigate arbitration proceedings. As will be seen later, a notice from either party to the other requesting the referral of a dispute to arbitration does not mean that arbitration proceedings will start immediately; however, the request can be made at any time.

The Article sets out the scope of the happenings to which arbitration can relate, as follows:

(a) The construction of the Contract, which means literally a dispute over interpretation of the meanings of the printed documents.

(b) Anything arising during the carrying out of the works.

(c) Any disputes arising over the exercise of the Architect's discretion where it is granted in a clause.

(d) The withholding of a certificate by the Architect to which the Contractor is entitled (this applies to any certificate, not just payment certificates).

(e) Adjustment of the Contract Sum, the final account.

(f) The rights and liabilities of both parties under Clause 27 determination.

(g) The same under Clause 28 – Determination.

(h) The same under Clause 32 – Hostilities.

(j) The same under Clause 33 – War damage.

(k) Either party withholding consent or agreement unreasonably where it is a pre-condition to some further action.

There are these specific exclusions to arbitration proceedings:

   (a) Under Clause 19A – Fair wages,
   (b) Under Clause 31.9 – Finance Act,
   (c) Under the VAT Agreement – Clause 3.

In all three cases above there are established statutory powers for dealing with disputes.

The Article envisages than an Arbitrator should be appointed who is mutually satisfactory to the parties. However, if this proves impossible to achieve within a 14-day period starting from the receipt of either party's request to proceed to arbitration, then it is agreed that either party can request the President or Vice-President of the Royal Institute of British Architects to appoint a suitable person. It is possible in either case to have an Arbitrator appointed by name before any proceedings take place, i.e., after Practical Completion.

5.1.4   This has been referred to as the 'enjoinder clause' which simply is a third party joining in an action of two other parties where the interests are the same.

With this concept in mind, the Article states that any dispute referred to arbitration which is common in its issues to a related dispute between

    the Employer/Nominated Sub-contractor,
or the Contractor/Nominated Sub-contractor,
or the Employer/Nominated Supplier,
or the Contractor/Nominated Supplier,

will, by this article, be agreed by the Employer and the Contractor to be settled by the same appointed Arbitrator.

Notwithstanding this very tidy concept of 'putting like disputes in the same basket', the Article follows on by giving the Employer the option to delete these enjoinder clauses, which he does by suitably striking out the entry in the Appendix. This has the effect of each dispute going to arbitration separately and could, of course, result in decisions over like disputes not being the same.

The matter is further complicated under Article 5.1.5 if either party considers that the Arbitrator originally appointed will not be suitable in a like dispute, making a further appointment of an Arbitrator necessary. Thus we might have the ludicrous situation of an arbitration to decide upon an Arbitrator for an arbitration.

5.2   With the exceptions following, all arbitration proceedings will take place only after either Practical Completion or termination of the works:

   (a) Both parties agree in writing to immediate arbitration.
   (b) Appointment of replacement Architect – Article 3.
   (c) Appointment of replacement Surveyor – Article 4.
   (d) Whether or not an instruction is empowered by the Conditions – Clause 4.

(e) Whether or not a certificate has been improperly withheld – several clauses involved here.

(f) Whether a certificate is not in accordance with the conditions – several clauses involved here.

(g) Whether or not the Contractor's objection is reasonable – Clause 4.1 which refers to Clause 13.1.2.

(h) Disputes over extensions of time – Clause 25.

(j) Disputes over hostilities – Clause 32

(k) Disputes over war damage – Clause 33.

It will be noticed that, with exception of the first point, all the other happenings listed are vital to whether or not the Contractor performs some action, which in turn will affect the progress of the works. Accordingly, the Contract recognizes the importance of having the matters settled as soon as possible.

5.3   With the following exceptions, the Arbitrator is given sweeping powers to review just about everything he considers necessary to enable him to reach a decision.

1   *Clause 4.2:* If the Contractor has requested that the Architect prove under which clause he is empowered to issue an instruction, and the Architect does so satisfactorily. The Contractor in turn must carry out that instruction, and by so doing he admits its validity.

2   *Clause 3.9:* There is a 14-day period following the issue of the Final Certificate during which either party can request arbitration over any of the matters contained in the Final Certificate. If this period ends without any arbitration being requested, then more of the matters which the Final Certificate makes conclusive evidence can be referred to arbitration. The opportunity to do so has now passed.

3   *Clauses 38.4.3, 39.5.3* and *40.5:* These are the three fluctuation clauses and the particular sub-clauses where in each clause the Quantity Surveyor and the Contractor can agree a lump sum with regard to any or all fluctuations. It is then this agreement that is not subject to arbitration.

5.4   Here the award of the Arbitrator is stated to be final and binding upon the parties.

The Arbitrator's conduct, bias, unreasonable time to get the proceedings started, or his handling of the proceedings themselves, could be suitable grounds for either party to appeal to the Courts. The award itself could be contested in the Courts on the grounds that it is biased, unclear, not consistent with the terms of the Contract or the dispute, or that an error had been made, or not all the evidence considered.

Remember that Arbitrators, although expert in their own field, will seldom also be legally trained. Accordingly, either party may appeal to the Courts on a point of law.

The ruling of a Court might set aside an award, change or modify it, or in extreme cases, dismiss the Arbitrator.

5.5   Here the law of England is to apply to all matters regardless of nationality, residence or domicile of either party, or the situation of the works. Also, the Arbitration Acts 1950 to 1979 shall apply to all arbitrations. Note that the location of the arbitration proceedings has no bearing.

By having a Scottish Supplement, Scotland automatically amends the contract for Scottish law.

In all other cases, if the law of England is not to apply, or the Arbitration Acts are not to apply, amendments will have to be made to Article 5.5 before any Contract is signed.

1. LOCAL AUTHORITIES EDITIONS: PRIVATE EDITIONS: DIFFERENCES

The principal differences between the Conditions of the two 'with Quantities' editions are:

1980

Clause	Local Authorities Edition	Private Edition
Generally	The expression 'Architect/Supervising Officer' is used.	The expression 'Architect' is used.
1.3	Definition of Supervising Officer given.	**No** definition of Supervising Officer given.
5.1	Contract documents shall remain in custody of the **Employer**.	Contract documents to remain in custody of the **Architect or the Quantity Surveyor**.
19A	Fair Wages clause	**No** provision
21.1.2	The **Employer** may require the Contractor to produce for inspection by the **Employer** evidence that insurances are properly maintained.	The **Architect** may require the Contractor to produce for inspection by the **Architect** evidence that the insurances are properly maintained.
21.2.2	The **Employer** approves insurers.	The **Architect** approves insurers.
22A.2	The **Employer** approves insurers.	The **Architect** approves insurers.
22A.3.1	The **Employer** may require the Contractor to produce for inspection by the **Employer** evidence that any 'all-risks policy' is properly endorsed and maintained.	The **Architect** may require the contractor to produce for inspection by the **Architect** evidence that any 'all-risks policy' is properly endorsed and maintained.
22B.1	**No** provision requiring Employer to insure against Clause 22 Perils, held at Employer's sole risk.	The Employer shall maintain a proper policy of insurance against Clause 22 Perils: if the Employer fails to show that he has done this, the contractor may so insure and recover costs: see footnote (n).
22C.1	**No** provision for the Employer to produce evidence that he has insured or for Contractor to insure.	If the Employer fails to show that he has maintained adequate insurance against Clause 22 Perils, the Contractor may so insure and recover costs.
27.3	The Employer is entitled to determine the employment of the Contractor if the Contractor has been engaged in the corrupt practices defined.	**No** provision.

1980 Clause	Local Authorities Edition	Private Edition
28.1.4	**No** provision.	The Contractor may determine his employment on the banktruptcy or liquidation of the Employer.
30.5.3	The Employer is trustee of the Retention but has **no** obligation to place the Retention in a separate trust account.	The Employer as trustee has to place the Retention in a separate trust account if the Contractor or any Nominated Sub-Contractor so requests.
30.6.2.15	**No** provision	Adjustment of Contract Sum to reflect Contractor's recovery for insuring under Clauses 22B.1 or 22C.1.
30.8	The Architect shall issue the final Certificate **within the** (variable) **period stated in Appendix II** from the latest of the three events listed.	The Architect shall issue the final Certificate within **3 months** from the latest of the three events listed (no provision for appendix entry).
40.1.3	Provision for Non-Adjustable Element under Price Adjustment Formulae to be stated in Appendix II.	**No** provision for Non-Adjustable Element under Price Fluctuations Formulae.

In the 1963 Private Editions, Clause 3(8) provided for Architect's certificates to be issued to the Contractor whereas the 1963 Local Authorities Editions provided for them to be issued to the employer (except the certificate related to Clause 27(d) (1963)). The 1963 Local Authorities Edition's provision has been applied to the 1980 Local Authorities and Private Editions (see Clause 5.8 (1980)), so this former difference between the two editions has been removed.

2. WITH QUANTITIES EDITIONS: WITHOUT QUANTITIES EDITIONS: DIFFERENCES

The principal differences between the Conditions of the two LOCAL AUTHORITIES editions are:

1980

Clause	With Quantities	Without Quantities
1, 2, 5, 6, 8, 9, 13, 14, 19, 21, 25, 26, 29, 30, 35, 36, 38, 39	The expressions 'Contract Bills' or 'Bills of Quantities' are used.	The expression 'Specification' is used.
5, 6, 13, 30, 38, 39	The expression 'Contract Bills' is used	The expression 'Schedule of Rates' is used.
5	The expression 'prices in the Contract Bills' is used.	The expression 'rates contained in the Schedule of Rates' is used.
1.3	Definition of 'Contract Bills' is given. No definition of 'Specification' is given	**No** definition of 'Contract Bills' is given. Definition of 'Specification' is given.
40	Fluctuations: Price Adjustment Formulae	**No** provision.

The essential differences occur in Clauses 1.3, 2.2, 5.3.1.2, 13 and 40. Similar differences occur between the PRIVATE WITH QUANTITIES and WITHOUT QUANTITIES editions.

Scotland

Clause 15 – Value added tax – supplemental provisions

Supplemental Provisions (the VAT Agreement)

Clause 19A – Fair wages

There are no changes to the above clauses for use in Scotland.

Arbitration

The Scottish Supplement Appendix I Part IV to the Scottish Building Contract sets out two Clauses, 41 and 42, for arbitration in Scotland.

As stated in the note under the title, Part IV is included in the Scottish Building Contract for convenience so that it may be included by reference when the formal Scottish Building Contract is not being executed. It is quite normal for a tender and letter of acceptance to form a sound basis for a building contract.

As a tender will incorporate the Conditions of Contract and the arbitration clauses are not included therein, this is the method of overcoming the difficulty. by making Clauses 41 and 42 amendments to the Conditions of Contract in the Supplement, Clauses 4 and 5 of the Scottish Building Contract are included in the Conditions. This is essential in the case of a contract formed by a tender and a letter of acceptance.

41   Although this clause is essentially the same as the matters in its English counterpart, the following significant differences arise:

(a) Should the parties fail, after 14 days, to agree on a mutual Arbiter, they agree to an Arbiter being appointed by the Sheriff of the Sheriffdom within which the works are situated.

(b) The enjoinder clause is not optional and there is no entry in Appendix II to be struck out.

(c) Contrary to the normal powers of an Arbiter, it is expressly written in that he has the power to award damages and expenses to or against any of the parties.

(d) The law of Scotland shall apply to all arbitrations and the award of the Arbiter being final and binding is qualified by Section 3 of the Administration of Justice (Scotland) Act 1972.

(e) It is expressly stated that the Arbiter will be entitled to remuneration and expenses.

42   The contract to be regarded as a Scottish Contract and the law of Scotland to prevail.

With regard to an appeal to the Courts over some aspect of arbitration, it should be pointed out that in a formal submission, i.e. a deed of submission, it is quite normal to include the appointment of a Clerk to the Arbitration. This clerk is usually a solicitor who can guide the Arbiter both in the procedures and on points of law. Both parties can channel questions through the Clerk to the Arbiter and the legal aspects can thus be effectively disposed of without recourse to the Courts.